Aliquot Cycles for Elliptic Curves with Complex Multiplication

Thomas Morrell

Washington University in Saint Louis

tmorrell@wustl.edu

21 March 2013

Abstract

We review the history of elliptic curves and show that it is possible to form a group law using the points on an elliptic curve over some field L. We review various methods for computing the order of this group when L is finite, including the complex multiplication method. We then define and examine the properties of elliptic pairs, lists, and cycles, which are related to the notions of amicable pairs and aliquot cycles for elliptic curves, defined by Silverman and Stange in [15]. We then use the properties of elliptic pairs to prove that aliquot cycles of length greater than two exist for elliptic curves with complex multiplication, contrary to an assertion of [15], proving that such cycles only occur for elliptic curves of j-invariant equal to zero, and they always have length six. We explore the connection between elliptic pairs and several other conjectures, and propose limitations on the lengths of elliptic lists.

Acknowledgments

I would like to thank my thesis advisor, Professor Matt Kerr, for his assistance in helping me learn about elliptic curves over the last year and a half, and for reviewing this paper. I am also indebted to Liljana Babinkostova and Boise State University for hosting me this past summer through an REU program (funded by NSF Grant DMS-1062857). Section 4.2 derives largely from work completed in Boise during the REU program (see [2]).

Contents

Chapter 1

Introduction

In [15], Silverman and Stange define an amicable pair for an elliptic curve $E/\mathbb{Q}$ to be a pair of primes (p, q) such that $\#\tilde{E}(\mathbb{F}_p) = q$ and $\#\tilde{E}(\mathbb{F}_q)$. In Section 4.2 (which follows work I completed in 2012 as part of an REU program, see [2]), we define the similar notion of an elliptic pair (Definition 4.2.1) for a set of elliptic curves E/L (where L is an extension field of $\mathbb{Q}$) with complex multiplication (CM) by an imaginary quadratic order $\mathfrak{o}_K$. We show that we can recreate many of the results from [15] using elliptic pairs, in addition to a few new results.

In [15], the notion of an aliquot cycle for E is introduced, in which a set of n primes $p_i \geq 5$ is defined such that $\#\tilde{E}(\mathbb{F}_{p_i}) = p_{i+1 \mod n}$ for all $1 \leq i \leq n$. We define an elliptic cycle (Definition 4.2.14) to be the analogue of an aliquot cycle for elliptic pairs. Silverman and Stange show that no aliquot cycles exist for elliptic curves with complex multiplication, except in the case that $j(E) = 0$. They predict that no aliquot cycles with $n \geq 3$ exist in this case, and they prove this for $n = 3$. We use elliptic cycles to find that an elliptic cycle exists of length $n = 6$, and then we use this result to find an aliquot cycle of length $n = 6$ for $E : y^2 = x^3 + 15$. We also show that cycles exist only for $n = 1, 2, 6$, and we place restrictions on the primes p which can be part of 1- or 6-cycles.

We also suggest a heuristic for determining the number of elliptic pairs below a given bound which utilizes the class number of K.

Chapter 2 introduces elliptic curves E, reviewing the historical motivation for the study of elliptic curves and defining "addition" of points on E. We show that this operation forms a group law, enabling the use of group theory in placing restrictions on the number of points on the reduction of E over a finite field. In Section 2.4, we define complex multiplication and list all the j-invariants of elliptic curves defined over $\mathbb{Q}$ that have CM by $\mathfrak{o}_K$.

Chapter 3 explores the theory of CM, with a focus on determining the number of points on the reduction of an elliptic curve with CM over a finite field. Section 3.1, we follow Chapter 13 of [8], in which Lang follows Max Deuring's 1941 paper "Die Typen der Multiplicatorenringe elliptischer Funktionenkörper" ("The types of multiplication rings of elliptic function fields"). We rederive Deuring's Reduction Theorem (Theorem 3.1.17), which forms the basis of the CM method of Atkin and Morain (see [1], or Section 3.2). We conclude Chapter 3 by considering the special cases of $j(E) = 0, 1728$.

In Chapter 4, we introduce Größencharakter and use them, along with the CM method, to prove our main results for elliptic pairs. Section 4.1, we closely follow Sections II.8-10 of [14] (with additional background) to define and prove some basic results of Größencharakter. We conclude the section with two Theorems involving Größencharakter from [15]. These results are used to prove Theorem 4.2.7 and Corollary 4.2.10.

There exists a correspondence between monic quadratic prime-generating polynomials f and elliptic lists (Definition 4.2.12). We use the properties of elliptic lists to place a limit on the number of consecutive values of these polynomials which can be prime. We also find that the density of primes represented by $f(n)$ ($n \in \mathbb{Z}$) should be zero (because the number of primes less than X represented by $f(n)$ is $O(\sqrt{X}/\log^2 X)$ - it is a fraction the number of elliptic primes (Definition 4.2.2) less than X, while the number of primes is $\Theta(X/\log X)$).

In cryptography, elliptic curves may be used to establish a key (for use in a symmetric key algorithm) via a protocol similar to the Diffie-Hellman key exchange. In order to use them, however, we have to be able to create curves which will maximize the security of the keys derived from them. If E has order m modulo p, then the relative security of a key derived from an unknown multiple of a point is no more than $\varphi(\varphi(m))$. Since this is in general maximized for prime m, and we are guaranteed that the point (so long as it is not the identity) does not have a smaller order in this case, we seek the ability to quickly generate elliptic curves of prime order. We can use elliptic pairs to find curves E of prime order over some large prime p (see [4] for an algorithm based on a similar idea). In order to improve upon existing algorithms, a better understanding of the distribution of elliptic pairs will be required, since it is relatively fast and easy to find a representative curve given an elliptic pair (see [4], [12]).

We find the first term in the (conjectured) asymptotic expansion for the number of elliptic pairs less than X. If $K = \mathbb{Q}(\sqrt{-d})$, where d is positive, square-free, and $d \equiv_8 3$, then the number of elliptic pairs less than X is asymptotic to $C\frac{\sqrt{d}}{h(-d)^2}\frac{X}{\log^2 X}$ for some positive constant $C \approx 0.16$ (see Subsection 4.2.4). In the future, an improved heuristic may enable the increased ability to single out elliptic primes given K. Normally, we seek K with class number at least 200 for added security, so it is important to be able to specify K in advance, which is not possible in [4].

Chapter 2

Basics of Elliptic Curves

In Chapter 2 we review the basics of elliptic curves, beginning with the historical motivation for their study (Section 2.1). We then define a group law (Section 2.2) enabling us to "add" points on an elliptic curve E over a given field L. We then use the group law in Section 2.3 to determine the number of points on an elliptic curve over a finite field via Schoof's algorithm (Algorithm 2.3.3). We conclude the chapter with some basic results about the theory of complex multiplication (Section 2.4), which will set the stage for the rest of the paper.

2.1 Historical Motivation for Elliptic Curves

Historically, elliptic curves arose from the study of elliptic integrals, which determine arc length on an ellipse. Let a be the semi-major axis, and let b be the semi-minor axis of an ellipse. Then the circumference of the ellipse is $4aE(\sqrt{1-(b/a)^2})$, where

$$E(k) = \int_0^{\pi/2} \sqrt{1 - k^2 \sin^2 \theta}\, d\theta = \int_0^1 y\, dx$$

and y comes from the elliptic curve $E : y^2 = (1-x^2)(1-k^2x^2)$ $(0 < k < 1)$ [13].

We say that two curves C_1 and C_2 are birationally equivalent if there exist rational functions transforming points on C_1 into points on C_2, and vice versa. Such curves share many similar features, regardless of the field over which they are studied. It is possible to show that E, above, is birationally equivalent to $E' : Y^2 = X(X-1)(X-\lambda)$, which in turn is birationally equivalent to $E'' : Y^2 = X^3 + AX + B$ (except in characteristic 2 and 3) [13].

If we try to compute $\int_\infty^x \frac{dt}{\sqrt{t(t-1)(t-\lambda)}}$, then we see that because the square root function branches in $\mathbb{C}$, the integral is path-dependent. However, it can be shown that the integral is unique up to the addition of $n_1\omega_1 + n_2\omega_2$ for some fixed $\omega_1, \omega_2 \in \mathbb{C}$ dependent upon E', with $n_1, n_2 \in \mathbb{Z}$ [13]. We define a lattice Λ to be the set of all $\mathbb{Z}$-linear combinations of ω_1 and ω_2.

This motivates the study of elliptic functions, which are meromorphic functions $f(z)$ on $\mathbb{C}$ which satisfy $f(z+\omega) = f(z)$ for all $\omega \in \Lambda$ for some lattice Λ, and for all $z \in \mathbb{C}$. If we fix Λ, then the set of elliptic functions is finitely generated. Let

$$\wp(z;\Lambda) = \frac{1}{z^2} + \sum_{\omega \in \Lambda, \omega \neq 0} \left[\frac{1}{(z-\omega)^2} - \frac{1}{\omega^2} \right]$$

be the Weierstraß $\wp$-function. Then we have the following Theorem (Theorem 3.2 of Chapter VI of [13]):

Theorem 2.1.1. *Every elliptic function over a fixed lattice Λ is a rational combination of $\wp(z;\Lambda)$ and $\wp'(z;\Lambda)$.*

It turns out that $(\wp')^2 = 4\wp^3 - g_2\wp - g_3$, where $g_2 = 60 \sum_{\omega \in \Lambda, \omega \neq 0} \omega^{-4}$ and $g_3 = 140 \sum_{\omega \in \Lambda, \omega \neq 0} \omega^{-6}$, the equation of another elliptic curve. This tells us that when considered over $\mathbb{C}$, elliptic curves have the nice property that they are isomorphic to a torus $\mathbb{C}/\Lambda$.

In general, we can multiply ω_1 and ω_2 by some complex unit without changing any of the features of the elliptic curve, although g_2 and g_3 will change. This enables us to construct an infinite number of elliptic curves which are isomorphic to one another, so long as we keep $\tau = \omega_1/\omega_2$ fixed. Given an elliptic curve $E : y^2 = x^3 + Ax + B$, it is often difficult to compute τ, however, so we need some mechanism for determining whether two elliptic curves E and E' are isomorphic. To this end, we define the j-invariant of an elliptic curve (or its corresponding lattice), which is fixed among all elliptic curves which are isomorphic to one another. Let

$$J(\tau) = \frac{g_2^3}{\Delta_\tau} = \frac{g_2^3(\tau)}{g_2^3(\tau) - 27g_3^2(\tau)}$$

and $j(\tau) = 1728J(\tau)$. Then

$$J(E) = \frac{4A^3}{4A^3 + 27B^2} = \frac{-64A^3}{\Delta_E}$$

and $j(E) = 1728J(E)$. The normalization j is more commonly used than J, due to the fact that the coefficients of its Fourier series are integral [5].

Note that j is infinite when $\Delta = 0$. In this case, E is called singular, and then it is not actually an elliptic curve. In this case, E either has a cusp or self-intersects at a point. As we will see in Section 2.3, E, as written above, will be singular no matter what the coefficients over certain fields. In this case (and others), we can write the more general

$$E : y^2 + a_1xy + a_3y = x^3 + a_2x^2 + a_4x + a_6$$

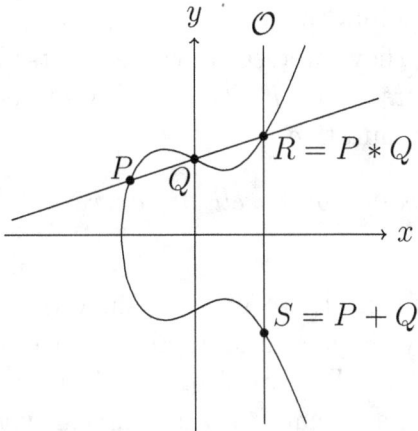

Figure 2.1: Point addition on $E : y^2 = x^3 - 2x + 4$.

to represent non-singular elliptic curves E over any field. Then if

$$b_2 = a_1^2 + 4a_2,$$
$$b_4 = a_1 a_3 + 2a_4,$$
$$b_6 = a_3^2 + 4a_6,$$
$$b_8 = a_1^2 a_6 + 4a_2 a_6 - a_1 a_3 a_4 + a_2 a_3^2 - a_4^2,$$
$$c_4 = b_2^2 - 24b_4,$$
$$c_6 = -b_2^3 + 36b_2 b_4 - 216b_6,$$

we can write

$$\Delta = -b_2^2 b_8 - 8b_4^3 - 27b_6^2 + 9b_2 b_4 b_6,$$
$$j = c_4^3/\Delta$$

[5]. This will come in particularly handy when E is over a field of characteristic $p = 2$ or 3.

2.2   The Group Law

Define the operation $* : E \times E \to E, P * Q \mapsto R$, where P, Q, R are collinear, and define $+ : E \times E \to E, P + Q \mapsto S$, where S is the reflection of R about the x-axis (see Figure 2.1). Alternatively, we can choose any point O on E, including a point at infinity, denoted by $\mathcal{O}$, when E is written in Weierstraß form, to be our "identity point" (the reason for this name will be elucidated when we prove that O is the identity for $+$). Then $P + Q = (P * Q) * O$. In the case that $O = \mathcal{O}$, we choose a vertical line through $P * Q$ and take the other point of intersection.

It is not obvious that $*$ (or $+$) is a well-defined operation (although $+$ follows from $*$), but this result follows from Bezout's theorem, which states that two projective curves of degrees

d_1 and d_2 intersect in $d_1 \cdot d_2$ points (including multiplicity). Since lines are of degree 1 and elliptic curves are of degree 3, they intersect in three points. If we fix P and Q, then there is exactly one other point of intersection, $P * Q$, so $*$ is well-defined. We see immediately that $*$ is commutative, so $+$ is commutative.

Theorem 2.2.1. *The set of points on an elliptic curve E forms an abelian group under the operation $+$ defined above.*

In order to prove Theorem 2.2.1, we have to show that $\forall P, Q \in E, P + Q \in E$, $+$ is associative, there is an identity element, and all points have inverses. That E is closed under $+$ is obvious from the definition. The proof of associativity is more difficult, so we save it for the end of the section. We have that $P * O$ is collinear with P and O, so the line through $P * O$ and O intersects E at P. Therefore $P + O = (P * O) * O = P$, so O is an identity element. Next, we claim that the inverse of P is $-P = P * (O * O)$. The line through P and $-P$ intersects E at $O * O$, and $(O * O) * O = O$, so $P + (-P) = O$. Note that to find $O * O$, we use the line tangent to E at O and take the third point of intersection. For lines through $\mathcal{O}$, we use a vertical line, and we set $\mathcal{O} * \mathcal{O} = \mathcal{O}$ (here, $\mathcal{O}$ is a **flex point** - i.e., a point P such that $P * P = P$).

When we define the addition of points, we usually take $O = \mathcal{O}$, and we follow this convention in Figure 2.1 and for the rest of the paper. In this case, if $P = (x_1, y_1)$, $Q = (x_2, y_2)$, $P * Q = (x_3, y_3)$, and E is in minimal Weierstraß form $E : y^2 = x^3 + Ax + B$, then setting $\lambda = \frac{y_2 - y_1}{x_2 - x_1}$, we find that $y_3 = y_1 + \lambda(x_3 - x_1)$. Then

$$y_3^2 = x_3^3 + Ax_3 + B$$
$$[y_1 + \lambda(x_3 - x_1)]^2 = x_3^3 + Ax_3 + B$$
$$0 = x_3^3 - \lambda^2 x_3^2 + (2\lambda(x_1 - y_1) + A)x_3 + (-y_1^2 + 2\lambda x_1 y_1 - \lambda^2 x_1^2 + B),$$

and we can eventually compute

$$x_3 = \lambda^2 - x_1 - x_2, \tag{2.1}$$
$$y_3 = y_1 + \lambda(x_3 - x_1) = y_2 + \lambda(x_3 - x_2). \tag{2.2}$$

Then, reflecting about the x-axis, we see that $P + Q = (x_3, -y_3)$.

Unfortunately, our definition for λ is undefined in the case that $P = Q$ (or $-Q$), so we cannot use (2.1) and (2.2). If $P = -Q$, then $P + Q = \mathcal{O}$, but if $P = Q$, then we have to find the slope of the tangent line to E at P (see Figure 2.2). Differentiating $y^2 = x^3 + Ax + B$ implicitly at P, we see that $2y_1 \, dy = (3x_1^2 + A) \, dx$, so

$$\lambda = \left. \frac{dy}{dx} \right|_P = \frac{3x_1^2 + A}{2y_1}.$$

The rest of the argument used before holds, so now we can use (2.1) and (2.2), with $x_1 = x_2$.

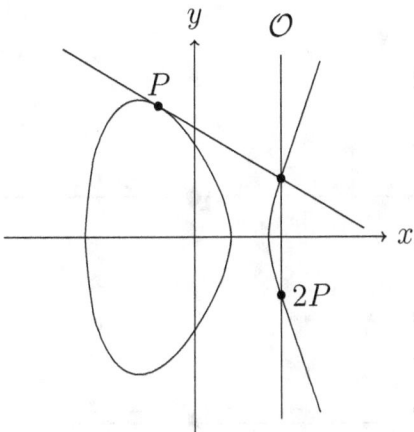

Figure 2.2: Point doubling on $E : y^2 = x^3 - 7x + 6$.

In order to compute the coordinates of multiples of points, we define the division polynomials ψ_m to be:

$$\psi_0 = 0$$
$$\psi_1 = 1$$
$$\psi_2 = 2y$$
$$\psi_3 = 3x^4 + 6Ax^2 + 12Bx - A^2$$
$$\psi_4 = 4y(x^6 + 5Ax^4 + 20Bx^3 - 5A^2x^2 - 4ABx - 8B^2 - A^3)$$

$$\vdots$$

$$\psi_{2m} = \frac{\psi_m}{2y} \cdot (\psi_{m+2}\psi_{m-1}^2 - \psi_{m-2}\psi_{m+1}^2) \qquad (m \geq 3)$$
$$\psi_{2m+1} = \psi_{m+2}\psi_m^3 - \psi_{m-1}\psi_{m+1}^3 \qquad (m \geq 2).$$

The division polynomials are such that

$$n(x, y) = \left(\frac{\phi_n(x)}{\psi_n^2(x)}, \frac{\omega_n(x, y)}{\psi_n^3(x, y)} \right),$$

where $\phi_n = x\psi_n^2 - \psi_{n+1}\psi_{n-1}$ and $\omega_n = \frac{1}{4y}(\psi_{n+2}\psi_{n-1}^2 - \psi_{n-2}\psi_{n+1}^2)$ [18]. Here,

$$nP = \underbrace{P + P + \ldots + P}_{n \text{ times}}.$$

The division polynomials will be especially useful in Section 2.3 when we discuss Schoof's algorithm.

We finish the section by proving associativity, and with it, finish the proof of Theorem 2.2.1.

Lemma 2.2.2. *If $P_1, ..., P_8$ are points in $\mathbb{P}^2$, no 4 on a line, and no 7 on a conic, then there is a unique 9th point Q such that any cubic through $P_1, ..., P_8$ also passes through Q.*

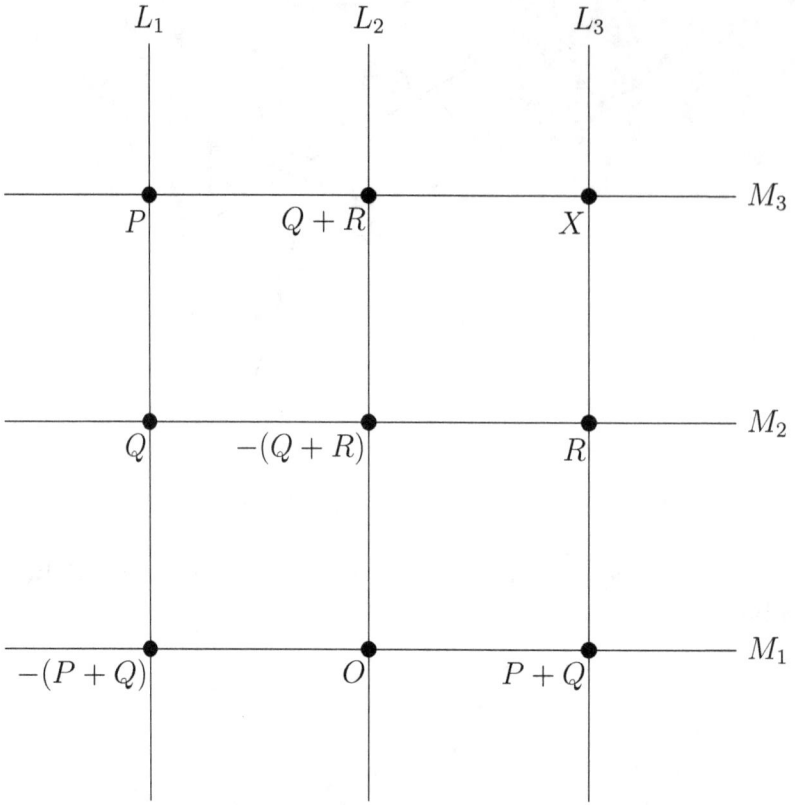

Figure 2.3: A geometric proof that addition on E is associative. E intersects the grid at the nine lattice points - and nowhere else. Therefore, $X = -(P + (Q + R)) = -((P + Q) + R)$, from which the fact that point addition on E is associative follows immediately.

Proof. This is a special case of the Cayley-Bacharach Theorem for two cubic curves (see, for example, Appendix A of [16]). □

Lemma 2.2.3. *As defined above, $+$ is associative on E.*

Proof. Note that Lemma 2.2.3 is trivial in the case that any of the points is O, or the case that all three points are the same. We shall prove associativity in the case that the three points being added are distinct and not the identity, leaving the case that two of the points are the same to the reader.

Note that points A, B, and $-(A + B)$ are collinear, so $P, Q, -(P + Q)$ all lie on a line, say L_1, and $P + Q, -(P + Q), O$ all lie on a line, say M_1. Also, $Q, R, -(Q + R)$ lie on the line M_2, and $Q + R, -(Q + R), O$ lie on L_2. Then $P + Q, R, -((P + Q) + R)$ lie on L_3 and $P, Q + R, -(P + (Q + R))$ lie on M_3. See Figure 2.3 for a visualization of this set-up.

Since E is a cubic, it intersects each of the six lines three times, and on lines L_1, L_2, M_1, M_2, we know that the points of intersection are the points we specified above, giving us 8 points. We have that $L_1 L_2 L_3 = 0$ and $M_1 M_2 M_3 = 0$ are both of degree three, so by Lemma 2.2.2, E intersects them in their ninth point of intersection, and nowhere else (assuming the conditions stated in the lemma, which we check below). Therefore, we must have that $-((P+Q)+R) = -(P+(Q+R))$, so $(P+Q)+R = P+(Q+R)$, as desired.

Now we check that no four points lie on a line, and no seven on a conic. By Bezout's Theorem, the intersection of any line with E contains 3 points, so no 4 points are collinear. Likewise, the intersection of any conic with E contains 6 points, so no 7 points are on the same conic. $\qquad\square$

2.3   Elliptic Curves over Finite Fields

In general, E is defined over some field, often $\mathbb{C}$ or $\mathbb{Q}$, with coefficients in the ring of integers of these fields. We can also study E over a field of characteristic other than zero, however. When we consider E over $\mathbb{F}_{p^n}$, where p is prime and $n \in \mathbb{Z}^+$, then we say that we are considering **the reduction of E over $\mathbb{F}_{p^n}$**, denoted by $\tilde{E}$. Points are found in the same way as before: we choose $x, y \in \mathbb{F}_{p^n}$ and check to see whether they satisfy the equation for E.

Since $\mathbb{F}_{p^n}$ has only a finite number of elements, it makes sense to ask how many points are on $\tilde{E}$. We denote this quantity by $\#\tilde{E}(\mathbb{F}_{p^n})$. We know that the point at infinity (now represented by $\tilde{\mathcal{O}}$) is a point on $\tilde{E}$, so $\#\tilde{E}(\mathbb{F}_{p^n}) \geq 1$, and there can be no more than 2 values of y for any given value of x such that $(x, y) \in \tilde{E}$, so $\#\tilde{E}(\mathbb{F}_{p^n}) \leq 2p^n + 1$. It seems unlikely that either of these extremes would ever be reached, since exactly half of the elements of $\mathbb{F}_{p^n}^\times$ are squares. We expect that $\#\tilde{E}(\mathbb{F}_{p^n}) = p^n + 1 + o(p^n)$ - but can we find a better bound for the error function? In fact, we can, as Hasse first did in 1936 (Theorem 2.3.1, below).

Thoerem 2.3.1. *Let $a = p^n + 1 - \#\tilde{E}(\mathbb{F}_{p^n})$. Then $|a| \leq 2p^{n/2}$. The interval $[p^n + 1 - 2p^{n/2}, p^n + 1 + 2p^{n/2}]$ is known as the Hasse interval.*

Proof. Let $q = p^n$ and let π_q be the qth-power Frobenius map:

$$\pi_q : \tilde{E} \to \tilde{E}, \qquad (x, y) \mapsto (x^q, y^q).$$

Since the Galois group $G_{\bar{\mathbb{F}}_q/\mathbb{F}_q}$ is topologically generated by π_q on $\bar{\mathbb{F}}_q$, we see that for a point $P \in \tilde{E}(\bar{\mathbb{F}}_q)$ that $P \in \mathbb{F}_q$ if and only if $\pi_q(P) = P$ [13]. Therefore, $\tilde{E}(\mathbb{F}_q) = \ker(1 - \pi_q)$ and $\#\tilde{E}(\mathbb{F}_q) = \#\ker(1 - \pi_q) = \deg(1 - \pi_q)$ (because $1 - \pi_q$ is separable). Because $\deg \pi_q = q$ and the degree map on $\text{End}(E)$ is a positive definite quadratic form [13], we can use Lemma 2.3.2 below to obtain the desired result.

$\qquad\square$

Lemma 2.3.2. *Let A be an abelian group and $d : E \to \mathbb{Z}$ be a positive definite quadratic form. Then for all $\psi, \phi \in A$,*

$$|d(\psi - \phi) - d(\phi) - d(\psi)| \leq 2\sqrt{d(\psi)d(\phi)}.$$

Proof. If $\psi = 0$, then the lemma is trivial, so set $\psi \neq 0$.

We see immediately that $L(\psi, \phi) = d(\psi - \phi) - d(\phi) - d(\psi)$ is bilinear. As d is positive definite,

$$0 \leq d(m\psi - n\phi) = m^2 d(\psi) + mn L(\psi, \phi) + n^2 d(\phi)$$

13

for integers m, n. We set $m = -L(\psi, \phi)$ and $n = 2d(\psi)$ to obtain

$$0 \leq d(\psi)[4d(\psi)d(\phi) - L(\psi, \phi)^2].$$

Since $\psi \neq 0$, $4d(\psi)d(\phi) \geq L(\psi, \phi)^2$. Taking square roots yields the desired inequality. $\qquad\square$

For $n = 1$, every possible value for a allowed by Theorem 2.3.1 is attained, although this is not true for $n > 1$. We note that $\pi_q^2 - a\pi_q + q = 0$ as an endomorphism of E [18], a result that will be of use in computing the order of $\tilde{E}$ later on.

2.3.1 Singular Reductions

Recall that the discriminant for the curve $E : y^2 = x^3 + Ax + B$ is $\Delta = -16(4A^3 + 27B^2)$. If $p | \Delta$, then E has bad reduction at p, that is, when we consider $\tilde{E}(\mathbb{F}_{p^n})$, we no longer have an elliptic curve. If $p \nmid \Delta$, then E has good reduction at p, so it is still an elliptic curve. It is important to note that the cases $p = 2, 3$ cause problems, so $E : y^2 = x^3 + Ax + B$ will not be an elliptic curve. We have to use the more general form from Section 2.1, instead. Due to this difficulty, for the rest of the paper, we assume that $p \neq 2, 3$, unless otherwise indicated. Note that all of the general results we prove can be used in characteristic 2 and 3.

It should be noted that if E has a singular reduction, we can still treat the points as a group and use the same group law as in Section 2.2, so long as we remove the singular point (there is exactly one). The type of group depends on the type of singular reduction. E has a cusp if $-c_6 \equiv_p 0$, a double point with tangents defined over $L = \mathbb{F}_{p^n}$ if $-c_6$ is a quadratic residue modulo p, or a double point with tangents not defined over L if $-c_6$ is not a quadratic residue modulo p. In the case that E has a cusp, the group of non-singular points G is isomorphic to the additive group $(L, +)$ and the order is $\#\tilde{E}(L) = p^n + 1$. In the case of a double point with tangents defined over L, $G \cong (L^*, \times)$ and $\#\tilde{E}(L) = p^n$. We say that E has split multiplicative reduction. In the remaining case, we say that E has non-split multiplicative reduction. Then G is isomorphic to the cyclic subgroup of $\mathbb{F}_{p^{2n}}$ of order $p^n + 1$, so $\#\tilde{E}(L) = p^n + 2$ [5]. We no longer have an elliptic curve in the case of singular reduction, however.

2.3.2 Schoof's Algorithm

Given the bounds on the order in Theorem 2.3.1, it is natural to try to determine the number of points on $\tilde{E}$ exactly. For singular curves, we saw that $\#\tilde{E}(\mathbb{F}_{p^n}) \in \{p^n, p^n + 1, p^n + 2\}$, with the order being determined by the type of bad reduction.

The fastest general-purpose algorithm for determining $\#\tilde{E}(\mathbb{F}_{p^n})$ is the Schoof-Elkies-Atkin (SEA) algorithm, an improvement upon Schoof's originial algorithm published in 1985 [9]. The complex multiplication method in Section 3.2 is faster, but it does not work for every E. SEA runs in time $\tilde{O}(\log^4 p^n)$, and Schoof's algorithm can be programmed to run in $\tilde{O}(\log^5 p^n)$, although the version we give here will run in $O(\log^8 p^n)$. Efficient implementations for $p > 2, n = 1$ and $p = 2, n \in \mathbb{Z}$ can be found in [10] and [11], respectively.

14

Schoof's algorithm is based on the idea that if $S = \{2, 3, 5, 7, ..., \mathcal{L}\}$ is a set of primes such that $\prod_{\ell \in S} \ell \geq 4\sqrt{p^n}$, then by the Chinese Remainder Theorem and Theorem 2.3.1 (which limits the number of orders E can take), we can determine a (from Theorem 2.3.1) uniquely by determining a modulo ℓ for all $\ell \in S$. For $\ell = 2$, note that $\#\tilde{E}$ is even (we include $\mathcal{O}$ in our count) if and only if $f(x) = x^3 + Ax + B$ has a root in $\mathbb{F}_{p^n}$. Recall that if $\deg f(x) \leq 3$, then $f(x)$ has a root in $\mathbb{F}_{p^n}$ if and only if $\gcd(x^{p^n} - x, f(x)) \neq 1$, so we can use the Euclidean algorithm to quickly determine a modulo 2.

For $\ell > 2$, note that because $\pi_q^2 - a\pi_q + q = 0$,

$$\left(x^{p^{2n}}, y^{p^{2n}}\right) + p^n(x, y) = a\left(x^{p^n}, y^{p^n}\right).$$

If (x, y) is a point of order ℓ (that is, $\ell(x, y) = \mathcal{O}$), then

$$(x', y') := \left(x^{p^{2n}}, y^{p^{2n}}\right) + (p^n)_\ell(x, y) = a\left(x^{p^n}, y^{p^n}\right),$$

where $(p^n)_\ell \equiv_\ell p^n$ and $|(p^n)_\ell| \leq \frac{\ell}{2}$. As (x^{p^n}, y^{p^n}) is a point of order ℓ, this relation determines a modulo ℓ.

Let $j(x, y) = (x_j, y_j)$ for integers j, where we can compute x_j and y_j using the division polynomials, as we did in Section 2.2. By equation (2.1),

$$x' = \left(\frac{y^{p^{2n}} - y_{q\ell}}{x^{p^{2n}} - x_{q\ell}}\right)^2 - x^{p^{2n}} - x_{q\ell}.$$

We seek to find j such that $(x', y') = (x_j^{p^n}, y_j^{p^n})$. Looking at the x-coordinates, we see that $(x', y') = \pm(x_j^{p^n}, y_j^{p^n})$ if and only if $x' = x_j^{p^n}$. In this case, $x' - x_j^{p^n} \equiv_{\psi_\ell} 0$ by the definition of the division polynomials. We then determine the sign by examining $Y = (y' - y_j^{p^n})/y$ modulo ψ_ℓ. If $Y \equiv 0$, then $a \equiv_\ell j$; otherwise, $a \equiv_\ell -j$ [18].

We must consider the case where $(x^{p^{2n}}, y^{p^{2n}}) = \pm p^n(x, y)$ (recall that $\ell(x, y) = \mathcal{O}$) because the x- and y-coordinates of the point at infinity are not well-defined, so we cannot define x' and y', as above.

Let $\psi_{p^n}^2(x, y) = \left(x^{p^{2n}}, y^{p^{2n}}\right) = p^n(x, y)$, so $a\pi_{p^n}(x, y) = \pi_{p^n}^2(x, y) + p^n(x, y) = 2p^n(x, y)$. Therefore, $a^2 p^n(x, y) = (2p^n)^2(x, y)$ and $a^2 \equiv_\ell 4p^n$. In particular, this can only be the case if p^n is a quadratic residue modulo ℓ, in which case we define w such that $w^2 = p^n$. Then

$$(\pi_{p^n} + w)(\pi_{p^n} - w)(x, y) = (\pi_{p^n}^2 - p^n)(x, y) = \mathcal{O},$$

so there exists a point P of order ℓ such that $\pi_{p^n} P = \pm wP$, which implies that

$$\mathcal{O} = (\pi_{p^n}^2 \mp a\pi_{p_n} + p^n)P = (p^n \mp aw + p^n)P.$$

Therefore, $a \equiv_\ell \pm 2w$. Since we are only considering one point of order ℓ and not all of them, we take a gcd between ψ_ℓ and the appropriate polynomials (analogous to those given before - see Algorithm 2.3.3) to determine which it is.

We are now ready to present Schoof's algorithm:

15

Algorithm 2.3.3. *This algorithm takes an elliptic curve $E : y^2 = x^3 + Ax + B$ over a finite field $L = \mathbb{F}_{p^n}$ and outputs the order $\#E(L)$.*

1. Choose a set of primes $S = \{2, 3, 5, ..., \mathcal{L}\}$ (with $p \notin S$) such that $\prod_{\ell \in S} \ell > 4\sqrt{p^n}$.

2. If $\ell = 2$, then $a \equiv_2 0$ if and only if $\gcd(x^3 + Ax + B, x^{p^n} - x) \neq 1$.

3. For each odd prime $\ell \in S$:

   [(a)] Let $q_\ell \equiv_\ell p^n$, with $|q_\ell| < \ell/2$.

   [(b)] Compute x' modulo ψ_ℓ, where

   $$(x', y') = \left(x^{p^{2n}}, y^{p^{2n}} \right) + q_\ell(x, y).$$

   [(c)] For $j = 1, 2, ..., (\ell - 1)/2$, do the following:

   [i.] Compute x_j, where $(x_j, y_j) = j(x, y)$.

   [ii.] If $x' - x_j^{p^n} \equiv_{\psi_\ell} 0$, go to step (iii). Otherwise, try the next value of j.

   [iii.] Compute y' and y_j. If $(y' - y_j^{p^n})/y \equiv_{\psi_\ell} 0$, then $a \equiv_\ell j$. Otherwise, $a \equiv_\ell -j$.

   [(d)] If all values of j have been tried without success, then if $\left(\frac{p^n}{\ell} \right) = -1$, $a \equiv_\ell 0$.

   [(e)] If $\left(\frac{p^n}{\ell} \right) = 1$, then let $w \equiv_\ell \sqrt{p^n}$, and compute

   $$g = \gcd(\text{numerator}(x^{p^n} - x_w), \psi_\ell).$$

   If $g = 1$, then $a \equiv_\ell 0$. If not, compute

   $$g' = \gcd(\text{numerator}((y^{p^n} - y_w)/y), \psi_\ell).$$

   If $g' = 1$, then $a \equiv_\ell -2w$. Otherwise, $a \equiv_\ell 2w$.

4. Use the Chinese Remainder Theorem and the values of a modulo ℓ to compute a modulo $\prod \ell$, and choose the value of a such that $|a| \leq 2\sqrt{p^n}$. Then $\#E(L) = p^n + 1 - a$.

We now consider the elliptic curve $E : y^2 = x^3 + 2x + 1$ over the field $L = \mathbb{F}_{5^2}$. We check the discriminant $\Delta_E = -944 \equiv_5 1 \not\equiv_5 0$ to make sure that E is not singular. We quickly compute $4\sqrt{5^2} = 20$, so we take $S = \{2, 3, 7\}$.

For $\ell = 2$, we compute $\gcd(x^3 + 2x + 1, x^{25} - x) = \gcd(x^3 + 2x + 1, 131x^2 - 706x - 347) = \gcd(131x^2 - 706x - 347, 578215x + 262143) = 1$, so $a \equiv_2 1$.

For $\ell = 3$, recall that $\psi_3 = 3x^4 + 12x^2 + 12x - 4$. We compute $q_\ell \equiv_3 25 \equiv_3 1$. Then $(x', y') = (x^{625}, y^{625}) + (x, y)$, so

$$x' = \left(\frac{y^{625} - y}{x^{625} - x} \right)^2 - x^{625} - x = (x^3 + 2x + 1) \left(\frac{(x^3 + 2x + 1)^{312} - 1}{x^{625} - x} \right)^2 - x^{625} - x.$$

But $x^{625} - x \equiv_{\psi_3, 5} 0$, so its multiplicative inverse does not exist. Therefore, a root of ψ_3 is defined over L, so $a \equiv_3 0$. Therefore, $a \equiv_6 3$.

For $\ell = 7$, we follow Schoof's algorithm to compute $a \equiv_7 5$, although we omit the computations here for the sake of brevity. Thus, $a \equiv_{42} 33$, so $a = -9$. Therefore, $\#E(L) = 25 + 1 - (-9) = 35$.

16

2.4 Complex Multiplication

Since the set of points on E form a group, we can look at the ring of endomorphisms of $(E, +)$. Over $\mathbb{C}$, it makes more sense to look at the torus $\mathbb{C}/\Lambda \cong E$. If we multiply points by an integer n, $n\Lambda \subset \Lambda$, so this is an endomorphism of E. For most curves, $\mathrm{End}(E) = \mathbb{Z}$, but some curves admit additional endomorphisms. Such curves are said to have **complex multiplication (CM)** because the additional endomorphisms are multiplications of points on Λ by complex numbers. We will see in Section 3.1 that if $\alpha \in \mathrm{End}(E)$ and $\alpha \notin \mathbb{Z}$, then $\alpha = a + b\sqrt{-d}$ for some square-free positive integer d.

As in Section II.2 of [14], take the curves

$$E : y^2 = x^3 + 4x^2 + 2x$$

and

$$E' : Y^2 = X^3 - 8X^2 + 8X,$$

both with $j = 8000$. Since $j(E) = j(E')$, $E' \cong E$, and in fact we have the isomorphism

$$\psi : E' \to E, \qquad (X, Y) \mapsto \left(-\frac{X}{2}, -\frac{Y}{2\sqrt{-2}} \right) =: (x, y).$$

We also have the isogeny (surjective homomorphism)

$$\phi : E \to E', \qquad (x, y) \mapsto \left(x + 4 + \frac{2}{x}, y\left(1 - \frac{2}{x^2}\right) \right) =: (X, Y).$$

From these, we have

$$\psi \circ \phi : (x, y) = \left(-\frac{1}{2}\left(x + 4 + \frac{2}{x} \right), -\frac{y}{2\sqrt{-2}}\left(1 - \frac{2}{x^2}\right) \right),$$

so

$$(\psi \circ \phi)^* \frac{dx}{y} = \frac{-\frac{dx}{2} + \frac{dx}{x^2}}{-\frac{y}{2\sqrt{2}}\left(1 - \frac{2}{x^2}\right)} = \sqrt{-2}\frac{dx}{y}.$$

Thus, E and E' have CM in $\mathbb{Q}(\sqrt{-2})$. We will work out an endormorphism α for elliptic curves with CM in $\mathbb{Q}(\sqrt{-1})$ in Section 3.3.

In general, we can only guarantee that elliptic curves with CM in $\mathbb{Q}(\sqrt{-d})$ exist with rational coefficients if the class number $h(-d) = 1$. This is because j is a rational function of the coefficients of the terms in the formula for E, and j is an algebraic integer of degree $h(-d)$ [14]. However, over finite fields in which $-d$ splits (i.e. E is not supersingular), we can define curves with CM in $\mathbb{Q}(\sqrt{-d})$ with coefficients in the field, regardless of how large $h(-d)$ is. This fact is not entirely obvious, and its proof goes beyond the scope of this work. See [12] for an explicit formula of a curve with CM in $\mathbb{Q}(\sqrt{-d})$.

There are 13 elliptic curves (up to isomorphism) over $\mathbb{Q}$ with CM. In Table 2.1, we give d, j, the conductor f of R (the order by which E has CM), and a representative curve $E/\mathbb{Q}$ with CM by the given order (see also Appendix A.3 of [14]).

Table 2.1: Representative Elliptic Curves over $\mathbb{Q}$ with CM (in $\mathbb{Q}(\sqrt{-d})$)

d	j	f	E
1	$1728 = 2^6 \cdot 3^3$	1	$y^2 = x^3 + x$
1	$287496 = 2^3 \cdot 3^3 \cdot 11^3$	2	$y^2 = x^3 - 11x + 14$
2	$8000 = 2^6 \cdot 5^3$	1	$y^2 = x^3 + 4x^2 + 2x$
3	0	1	$y^2 + y = x^3$
3	$54000 = 2^4 \cdot 3^3 \cdot 5^3$	2	$y^2 = x^3 - 15x + 22$
3	$-12288000 = -2^{15} \cdot 3 \cdot 5^3$	3	$y^2 + y = x^3 - 30x + 63$
7	$-3375 = -3^3 \cdot 5^3$	1	$y^2 + xy = x^3 - x^2 - 2x - 1$
7	$16581375 = 3^3 \cdot 5^3 \cdot 17^3$	2	$y^2 = x^3 - 595x + 5586$
11	$-32768 = -2^{15}$	1	$y^2 + y = x^3 - x^2 - 7x + 10$
19	$-884736 = -2^{15} \cdot 3^3$	1	$y^2 + y = x^3 - 38x + 90$
43	$-884736000 = -2^{18} \cdot 3^3 \cdot 5^3$	1	$y^2 + y = x^3 - 860x + 9707$
67	$-147197952000 = -2^{15} \cdot 3^3 \cdot 5^3 \cdot 11^3$	1	$y^2 + y = x^3 - 7370x + 243528$
163	$-262537412640768000 = -2^{18} \cdot 3^3 \cdot 5^3 \cdot 23^3 \cdot 29^3$	1	$y^2 + y = x^3 - 2174420x + 1234136692$

Chapter 3

The Complex Multiplication Method

In this chapter, we derive Deuring's Reduction Theorem (Theorem 3.1.17) in Section 3.1. We then use this result to find a formula for the number of points on an elliptic curve $E/\mathbb{F}_p$ with CM in $\mathbb{Q}(\sqrt{-d})$ in Section 3.2 (see Equation (3.1)). Finally, we consider the cases where $j(E) = 0, 1728$ in Section 3.3.

3.1 ℓ-adic Methods of Deuring

We closely follow Lang's exposition of Deuring's 1941 work "Die Typen der Multiplicatorenringe elliptischer Funktionenkörper" from Chapter 13 of [8].

Either there are no points of order p on an elliptic curve E considered in a field of characteristic $p > 0$, or the points of order p form a cyclic group $\mathbb{Z}/p\mathbb{Z}$ [8]. In the first of these cases, we say that E is **supersingular**, and in the latter, we say that E is **singular** or **generic**, depending on whether the j-invariant of E is transcendental over the chosen field.

Using ℓ-adic and p-adic representations, Deuring determined what happens to the endomorphism ring of an elliptic curve under reduction modulo p. Lang modifies Deuring's approach slightly to be better suited to modern notational preferences.

3.1.1 The ℓ-adic spaces

Let $E = E/\mathbb{F}$ be an elliptic curve, and let $\mathbb{F}$ have characteristic p. We consider points of E in a fixed algebraic closure $\overline{\mathbb{F}}$.

Definition 3.1.1. *For prime number ℓ, the ℓ-adic module $T_\ell(E)$ (also called the **Tate module**) is the set of infinite vectors*

$$(a_1, a_2, ...)$$

with $a_i \in E_{\ell^i}$ (so $\ell^i a_i = 0$) and $\ell a_{i+1} = a_i$.

We define addition componentwise such that $T_\ell(E)$ is a group. Likewise, multiplication by an ℓ-adic number is defined componentwise, where we approximate the ℓ-adic number by an integer modulo ℓ^i and multiply the ith component by this integer to get the new component for all $i \in \mathbb{N}$. We will denote the set of ℓ-adic integers by $\mathbb{Z}_\ell$, and we will use $\mathbb{Z}/\ell\mathbb{Z}$ (or $\mathbb{F}_\ell$ because ℓ is prime and we have a field) for the ring of integers modulo ℓ.

Theorem 3.1.2. *If $\ell \neq p$, then $T_\ell(E)$ is a free $\mathbb{Z}_\ell$-module of dimension 2. On the other hand, $T_p(E) = 0$, or is a free module of dimension 1 over $\mathbb{Z}_p$, according as we are in the supersingular or singular case.*

Proof. First let $\ell \neq p$, and let $x_1, x_2 \in T_\ell(E)$ have first components $a_{1,1}, a_{2,1}$, respectively, which are linearly independent over the field $\mathbb{Z}/\ell\mathbb{Z}$. Then x_1, x_2 are linearly independent over $\mathbb{Z}_\ell$ because otherwise the hypothesis on their first components could not be satisfied.

We shall use induction to prove that x_1, x_2 form a basis of $T_\ell(E)$ over $\mathbb{Z}_\ell$. Suppose that $\forall w \in T_\ell(E)$, we can write

$$w \equiv z_1 x_1 + z_2 x_2 \quad \mod \ell^n T_\ell(E)$$

for some $z_1, z_2 \in \mathbb{Z}$. We let $w = (b_1, ..., b_n, b_{n+1}, ...)$, so that

$$z_1(a_{1,1}, ..., a_{1,n+1}) + z_2(a_{2,1}, ..., a_{2,n+1}) = (b_1, ..., b_{n+1}) + (0, ..., 0, c_{n+1})$$

for some point c_{n+1} of order ℓ. By our choice of x_1, x_2, $\exists d_1, d_2 \in \mathbb{Z}$ such that

$$c_{n+1} = d_1 \ell^n a_{1,n+1} + d_2 \ell^n a_{2,n+1}.$$

Replacing $z_i \leftarrow z_i + d_i \ell^n$ extends the congruence

$$w \equiv z_1 x_1 + z_2 x_2 \quad \mod \ell^k T_\ell(E)$$

from $k = n$ to $k = n+1$. As the case $k = 1$ was assumed, this completes the proof for $\ell \neq p$.

Now we let $\ell = p$ and see that the set of points on E of order p^i is cyclic of order p^i in the singular case, so $T_p(E)$ is free over $\mathbb{Z}_p$. If there are no points of order p, then $T_p(E) = 0$. $\square$

As the $\ell = p$ case is relatively uninteresting, for the remainder of this section, we will assume that $\ell \neq p$.

If $\lambda : E \to E'$ is a homomorphism of elliptic curves, then λ induces a homomorphism

$$\lambda : T_\ell(E) \to T_\ell(E'),$$

with a similar result for T_p. Then

$$\lambda(a_1, a_2, ...) = (\lambda a_1, \lambda a_2, ...).$$

Theorem 3.1.3. *If endomorphisms $\lambda_1, ..., \lambda_r$ of E are linearly independent over $\mathbb{Z}$, then as endomorphisms of $T_\ell(E)$, they are linearly independent over $\mathbb{Z}_\ell$.*

Proof. Find $c_1, ..., c_r \in \mathbb{Z}_\ell$ such that $c_1\lambda_1 + \cdots + c_r\lambda_r = 0$. It suffices to prove that $\forall i, \ell | c_i$, because then we can divide by ℓ as needed to reach a contradiction (unless all $c_i = 0$). Let $c_i = m_i + \ell d_i$, with $m_i \in \mathbb{Z}$ and $d_i \in \mathbb{Z}_\ell$. It will suffice to show that $\forall i, \ell | m_i$ (reducing to the previous scenario).

Set $\lambda = m_1\lambda_1 + \cdots m_r\lambda_r \in \text{End}(E)$. Then because $0 = c_1\lambda_1 + \cdots + c_r\lambda_r = (m_1 + \ell d_1)\lambda_1 + \cdots (m_r + \ell d_r)\lambda_r$, we have that $\lambda = -\ell(d_1\lambda_1 + \cdots + d_r\lambda_r)$. If δ is the identity endomorphism, then this shows that λ factors through $\ell\delta$, and so $\exists \alpha \in \text{Emd}(E), \lambda = \ell\alpha$. Since $\lambda_1, ..., \lambda_r$ generate the space $\mathbb{Q}\lambda_1 + \cdots + \mathbb{Q}\lambda_r$ over $\mathbb{Q}$,

$$(\mathbb{Q}\lambda_1 + \cdots + \mathbb{Q}\lambda_r) \cap \text{End}(E)$$

is a lattice of rank r in $\text{End}(E) < \mathbb{Q}$. We know that $\alpha \in \mathbb{Z}\lambda_1 + \cdots \mathbb{Z}\lambda_r$, so $\forall i, \ell | m_i$, as desired. $\square$

By Theorem 3.1.3, we obtain the injection

$$\mathbb{Z}_\ell \otimes_\mathbb{Z} \text{End}(E) \to \text{End}_{\mathbb{Z}_\ell}(T_\ell(E));$$

we can see that our representation of $\text{End}(E)$ on $T_\ell(E)$ corresponds to tensoring with $\mathbb{Z}_\ell$.

We let $V_\ell(E) \supset T_\ell(E)$ with the first component a point on E of order a power of ℓ (let us define the set of such points by $E^{(\ell)}$). One can see that

$$V_\ell \cong \mathbb{Q}_\ell \otimes_{\mathbb{Z}_\ell} T_\ell,$$

and in fact for any $x \in V_\ell$, we can find s such that $\ell^s x$ has first component 0. Note that

$$0 \to T_\ell(E) \to V_\ell(E) \to E^{(\ell)} \to 0$$

is an exact sequence, with the mapping on the right being projection onto the first component.

For arbitrary ℓ,

$$\mathbb{Q} \otimes_\mathbb{Z} \text{End}(E) = \text{End}(E)_\mathbb{Q} \to \text{End}_{\mathbb{Q}_\ell}(V_\ell)$$

is a faithful representation. Since $\dim_{\mathbb{Q}_\ell}(V_\ell) = 2$, $\dim_{\mathbb{Q}_\ell} \text{End}_{\mathbb{Q}_\ell}(V_\ell) = 4$. This proves the following theorem:

Theorem 3.1.4. *In any characteristic,* $\dim_\mathbb{Q} \text{End}(E)_\mathbb{Q} \leq 4$ *and* $\dim_\mathbb{Z} \text{End}(E) \leq 4$.

The dimension is either 1, 2 (commutative), or 4 (not commutative).

3.1.2 Representations in Characteristic p

Proposition 3.1.5. *Let* $\alpha \in \text{End}(E)$ *be a non-trivial endomorphism. Then* $\mathbb{Q}(\alpha)$ *is quadratic imaginary.*

Proof. As $\mathbb{Q}(\alpha)$ is a commutative subfield of a division algebra of dimension 4 over $\mathbb{Q}$, it follows that $[\mathbb{Q}(\alpha) : \mathbb{Q}] = 2$, so α is quadratic (because α is not just a multiplication-by-n endomorphism when E has CM). In fact, α must be an automorphism of E other than the identity. Since the only real automorphisms of E are ± 1, α must be complex, and $\mathbb{Q}(\alpha)$ must be imaginary. $\square$

Remark 3.1.6. *For notational convenience, if $\lambda : E \to E'$ is a homomorphism, then we will represent its degree by $\nu(\lambda)$. We list without proof some of the properties of $\nu(\lambda)$ when $\lambda : E \to E$ is an endomorphism:*

  i. *if $E \cong \mathbb{C}/\Lambda$, then $\nu(\lambda) = (\Lambda : \lambda\Lambda)$ is the degree map for λ,*

  ii. *there exists an involution (map which is it's own inverse) of endomorphisms $\lambda \mapsto \lambda'$ with the property that $\lambda\lambda' = \lambda'\lambda = \nu(\lambda)\delta$, where δ is the identity map,*

  iii. *if $\mathbb{Q}(\lambda)$ is quadratic imaginary, as it is in Proposition 3.1.5, then $\lambda' = \nu(\lambda)\lambda^{-1}$ is the complex conjugate of λ, and $\nu(\lambda)$ is the norm of λ.*

Theorem 3.1.7. *Let E be defined over a finite field with $q = p^r$ elements, and let π_q be its Frobenius endomorphism. If $\pi_q \in \mathbb{Z}$, then $T_p = 0$. So if $T_p \neq 0$, then π_q is a non-trivial endomorphism.*

Proof. The degree of $\pi_q = q$. Assume that $\pi_q = n\delta$, so $p^r = q = \nu(\pi_q) = n^2$, and thus $n = p^m$ for some $m \in \mathbb{Z}$. Since $\mathbb{F}_q$ is finite, π_q is purely inseparable, so $p^m\delta$ has kernel 0. Therefore, $T_p = 0$. $\qquad\qquad\square$

Theorem 3.1.8. *Let E be an elliptic curve over a finite field $\mathbb{F}$ of characteristic p, and assume that $T_p(E) \neq 0$. Then:*

  i. $\mathrm{End}(E)_{\mathbb{Q}} = K$ *is a quadratic imaginary field, and* $\mathrm{End}(E) = \mathfrak{o}$ *is an order in K ($\mathfrak{o}$ is a $\mathbb{Z}$-lattice such that $K = \mathbb{Q}\mathfrak{o}$).*

  ii. *The prime p does not divide the conductor c of $\mathfrak{o}$.*

  iii. *The prime p splits completely in K.*

Proof. By Theorem 3.1.7, π_q is a non-trivial endomorphism of E. Since the representation of $\mathrm{End}(E)$ on T_p is faithful, it gives rise to an embedding of $\mathrm{End}(E)$ in $\mathbb{Z}_p$, so $\mathrm{End}(E)$ is commutative. Therefore, we know that $\mathrm{End}(E)$ has dimension two over $\mathbb{Z}$. By Proposition 3.1.5, $K = \mathrm{End}(E)_{\mathbb{Q}}$ is a quadratic imaginary field. Because K can be embedded in $\mathbb{Q}_p$, p must split completely in K.

It remains to show that $p \nmid c$. Since $\mathbb{Z} \subset \mathrm{End}(E)$, we know that $\mathfrak{o} = \mathbb{Z} + c\mathfrak{o}_K$. There exists $m \in \mathbb{Z}$ such that $\pi_q = m + c\alpha$ and $\pi_q' = m + c\alpha'$ for some $\alpha \in \mathfrak{o}_K$. Then $q\delta = \pi_q\pi_q' \equiv_{c\mathfrak{o}_K} m^2$. If we try to embed $\mathfrak{o}_K$ in $\mathbb{Z}_p$, then we find that $p|m$, so π_q kills the points of order p on E. But this contradicts the fact that π_q is purely inseparable, so $p \nmid m$, and thus $p \nmid c$. $\qquad\square$

Corollary 3.1.9. *Let $q = p^r$ be the number of elements of $\mathbb{F}$, and let $\pi = \pi_q$ be the Frobenius endomorphism. If $p\mathfrak{o} = \mathfrak{p}\mathfrak{p}'$ is the factorization of p in $\mathfrak{o} = \mathrm{End}(E)$, then $\pi\mathfrak{o} = \mathfrak{p}^r$ or $\pi\mathfrak{o} = (\mathfrak{p}')^r$, and any other generator of $\pi\mathfrak{o}$ is $\pm\pi$.*

Proof. As $\pi\pi' = q\delta$, in the unique factorization in $\mathfrak{o}$, only divisors of p can occur as divisors of π and π'. As $p \nmid \pi$, it follows that $\exists m \in \mathbb{N}$ such that (after permuting $\mathfrak{p}$ and $\mathfrak{p}'$, as necessary)

$$\pi\mathfrak{o} = \mathfrak{p}^m \quad \text{and} \quad \pi'\mathfrak{o} = (\mathfrak{p}')^m.$$

22

Thus, $\pi\pi'\mathbf{o} = p^r\mathbf{o}$, and $m = r$. Since E is not supersingular, the only automorphisms are $\pm\delta$ (the fourth or sixth roots of unity are automorphisms for $j_E = 1728$ and $j_E = 0$, respectively), proving the corollary. $\qquad\square$

Now we consider the case that E is supersingular, so $T_p(E) = 0$. If E' is isogenous to E, then $T_p(E') = 0$ as well. Since we will not need the next two results (we focus primarily on the case where p splits in K), we state them without proof.

Theorem 3.1.10. *Let $E/\mathbb{F}_q$ be an elliptic curve, with $q = p^r$. If $T_p(E) = 0$, then $j_E = j_E^{p^2}$.*

We see that $j_E \in \mathbb{F}_{p^2}$ if $T_p(E) = 0$, and therefore, there exist only finitely many isomorphism classes of elliptic curves E in characteristic p such that $T_p(E) = 0$.

Corollary 3.1.11. *Assume that E is supersingular, with invariant j, and that E is defined over $\mathbb{F}_p(j) = \mathbb{F}$. Then for $p \neq 2, 3$ we have:*

$$\pi_p^2 = -p\delta \qquad \text{if } j \in \mathbb{F}_p,$$

$$\pi_{p^2} = \pm p\delta \qquad \text{if } j \notin \mathbb{F}_p.$$

The formulas are similar for characteristic 2 or 3, but we are not interested in such cases, so we neglect these cases.

3.1.3   Representations and Isogenies

Recall that an **isogeny** is a surjective homomorphism between two elliptic curves E and E'.

Theorem 3.1.12. *Let $\lambda : E \to E'$ be an isogeny and $\nu(\lambda) = p^r$. The map $\mathrm{End}(E) \ni \alpha \mapsto \lambda\alpha\lambda^{-1} \in \mathrm{End}(E')$ is an isomorphism between $\mathrm{End}(E)$ and $\mathrm{End}(E')$.*

Proof. It will suffice to prove the theorem in the case that $r = 1$ because we can decompose λ into r isogenies, each of degree p. Furthermore, it suffices to show that for $\alpha \in \mathrm{End}(E)$, $\lambda\alpha\lambda^{-1} \in \mathrm{End}(E')$ because we then have the inverse map

$$\lambda'\lambda\alpha\lambda^{-1}\lambda'^{-1} = p\alpha p^{-1} = \alpha.$$

We first prove the case in which λ is separable. Then $\lambda\lambda' = p\delta$, so λ' is purely inseparable, and $\lambda^{-1} = p^{-1}\lambda'$. Suppose $\lambda\alpha\lambda^{-1} = \frac{1}{p}\beta$ for some $\beta \in \mathrm{End}(E)$. Then $\beta = \lambda\alpha\lambda'$. Since λ' is purely inseparable and λ is separable, $\ker\beta$ contains a point of period p. Hence, $\beta = p\gamma$ for some $\gamma \in \mathrm{End}(E')$, so $\gamma = \lambda\alpha\lambda^{-1}$, proving the theorem for this case.

If λ is purely inseparable, then there exists an isomorphism ε such that $\lambda = \varepsilon\pi$. Then $\lambda^{-1} = \pi^{-1}\varepsilon^{-1}$, so $\lambda\alpha\lambda^{-1} = \varepsilon\pi\alpha\pi^{-1}\varepsilon^{-1}$. For each point $x \in \pi(E)$, we have that

$$\pi\alpha\pi^{-1}(x) = \pi(\alpha(x^{1/p})) = \alpha^{(p)}(x),$$

where $\alpha^{(p)}$ is the image of α under the automorphism $c \mapsto c^p$ of the universal domain. Therefore, $\pi\alpha\pi^{-1} = \alpha^{(p)} \in \mathrm{End}(\pi E)$, from which we find that $\varepsilon\pi\alpha\pi^{-1}\varepsilon^{-1} \in \mathrm{End}(E')$. $\qquad\square$

We want to see how the modules T_ℓ correspond under isogenies; we will see that they act similarly to lattices Λ in $\mathbb{C}$. If we let $\lambda : E \to E'$ be an isogeny, then we can find its inverse $\lambda^{-1} \in \mathrm{Hom}(E', E)_\mathbb{Q}$, the tensor of $\mathrm{Hom}(E', E)$ with $\mathbb{Q}$.

Definition 3.1.13. *Let R be a commutative ring, and let M be an R-module. If $S \subset R$ is multiplicatively closed, then the **localization** of M with respect to S (denoted by $S^{-1}M$) is the set of equivalence classes of ordered pairs $(m, s) \in M \times S$ such that (m, s) and (n, t) are considered equivalent if $\exists u \in S$ such that $u(sn - tm) = 0$.*

Lemma 3.1.14. *Let S_ℓ be the multiplicative monoid of positive integers prime to ℓ, let $\mathfrak{o} = \mathrm{End}(E)$, and let $\mathfrak{o}_{(\ell)} = S_\ell^{-1}\mathfrak{o}$ be the localization of $\mathfrak{o}$ at ℓ. Let $\alpha \in \mathrm{End}(E)_\mathbb{Q}$. Then $\alpha T_\ell \subset T_\ell$ if and only if $\alpha \in \mathfrak{o}_{(\ell)}$.*

Proof. Let $\alpha \in \mathfrak{o}_{(\ell)}$. Then $\alpha T_\ell \subset T_\ell$. Conversely, if $\alpha T_\ell \subset T_\ell$, then $\exists \lambda \in \mathfrak{o}$ such that $m\ell^r \alpha = \lambda$ for some $m \in \mathbb{Z}$ prime to ℓ. Then

$$m\ell^r \alpha T_\ell \subset \ell^r T_\ell,$$

so $\lambda T_\ell \subset \ell^r T_\ell$ and $\lambda = \ell^r \beta$ for some $\beta \in \mathfrak{o}$. Therefore, $m\ell^r \alpha = \ell^r \beta$, so $m\alpha = \beta$, and thus $\alpha \in S_\ell^{-1}\mathfrak{o}$. So α is ℓ-integral, as desired. $\qquad\square$

Lemma 3.1.15. *Let $\lambda : E \to E'$ be an isogeny, and let M_ℓ be the set of vectors $(a_0, a_1, ...)$ in $V_\ell(E)$ such that $a_0 \in \ker \lambda$. Then $\lambda M_\ell = T_\ell(E')$.*

Theorem 3.1.16. *Let $\lambda : E \to E'$ be an isogeny, and let $\alpha \in \mathrm{End}(E)_\mathbb{Q}$. Let M_ℓ be the inverse image of $T_\ell(E')$ in $V_\ell(E)$ under λ. We have $\lambda \alpha \lambda^{-1} \in \mathrm{End}(E')$ if and only if $\alpha M_\ell \subset M_\ell$ for all ℓ.*

Proof. Let $p | \nu(\lambda)$. We can decompose λ into a product of an isogeny whose degree is prime to p, and an isogeny whose degree is p^r for some r. Using Theorem 3.1.12, we see that the theorem follows immediately in this case.

Now assume that $p \nmid \nu(\lambda)$, and suppose that $\lambda \alpha \lambda^{-1} \in \mathrm{End}(E')$. Then $\forall \ell, \alpha M_\ell = \lambda^{-1} \lambda \alpha \lambda^{-1} \lambda M_\ell \subset \lambda^{-1} T_\ell(E') \subset M_\ell$. Conversely, suppose that $\forall \ell, \alpha M_\ell \subset M_\ell$, so

$$\lambda \alpha \lambda^{-1} T_\ell(E') = \lambda \alpha M_\ell \subset \lambda M_\ell \subset T_\ell(E').$$

By Lemma 3.1.14, $\lambda \alpha \lambda^{-1}$ is ℓ-integral for each ℓ. It remains to show that $\lambda \alpha \lambda^{-1}$ is also p-integral. Suppose $\lambda \alpha \lambda^{-1} = p^{-r}\beta$ for some $\beta \in \mathrm{End}(E')$, and let $n = \nu(\lambda)$. Then $n\beta = p^r \lambda \alpha n \lambda^{-1} = p^r \gamma$ for some $\gamma \in \mathrm{End}(E')$, so $\lambda \alpha \lambda^{-1} = \frac{1}{n}\gamma$. For all $\ell | n$, $\frac{1}{n}\gamma T_\ell(E') \subset T_\ell(E')$, and $\gamma = m\gamma'$ for some $\gamma' \in \mathrm{End}(E')$ (by Lemma 3.1.14). Therefore, $\lambda \alpha \lambda^{-1} = \gamma' \in \mathrm{End}(E')$, proving our Theorem in the case $p \nmid \nu(\lambda)$, and thus overall. $\qquad\square$

3.1.4 Reduction of the Ring of Endomorphisms

We now examine how the ring of endomorphism reduces when we switch from examining elliptic curves in characteristic 0 to curves in characteristic p.

Let E be an elliptic curve defined over a number field, and let $\mathfrak{P}$ be a place (equivalence class of absolute values) of $\overline{\mathbb{Q}}$ with values in $\overline{\mathbb{F}}_p$. The place induces a discrete valuation ring. We have an isomorphism

$$E^{(\ell)} \to \tilde{E}_p^{(\ell)},$$

where $E^{(\ell)}$ denotes the group of points of $E/\overline{\mathbb{Q}}$ whose order is a power of ℓ, and $\tilde{E}_p = E(\mathfrak{P})$ is the place of E over p. Consequently we have the isomorphism

$$T_\ell(E) \to T_\ell(\bar{E}).$$

Although not an isomorphism,

$$T_p(E) \to T_p(\tilde{E}_p)$$

is a homomorphism. If $T_p(\bar{E}) \neq 0$, then the kernel of this map is a 1-dimensional module over $\mathbb{Z}_p$.

We now state Deuring's Reduction Theorem [DRT], which will be useful in determining the order of the group of points on E.

Theorem 3.1.17 (Deuring's Reduction Theorem). *Let E be an elliptic curve over a number field, with $\mathrm{End}(E) \cong \mathfrak{o}$, where $\mathfrak{o}$ is an order in an imaginary quadratic field K. Let $\mathfrak{P}$ be a place of $\overline{\mathbb{Q}}$ over a prime number p, where E has non-degenerate reduction $\tilde{E}_p$. The curve $\tilde{E}_p$ is supersingular if and only if p has only one prime of K above it (p ramifies or remains prime in K). Suppose that p splits completely in K. Let c be the conductor of $\mathfrak{o}$, and write $c = p^r c_0$, where $p \nmid c_0$. Then:*

(i) $\mathrm{End}(\tilde{E}_p) = \mathbb{Z} + c_0 \mathfrak{o}_K$ is the order in K with conductor c_0.

(ii) If $p \nmid c$, then the map $\lambda \mapsto \lambda(\mathfrak{P})$ is an isomorphism of $\mathrm{End}(E)$ onto $\mathrm{End}(\tilde{E}_p)$.

Proof. We will prove the theorem in the case that p splits in K, as this is the only case needed in future sections. For a complete proof, see [8].

Let p split in K: $p\mathfrak{o}_K = \mathfrak{p}\mathfrak{p}'$ ($\mathfrak{p} \neq \mathfrak{p}'$, $\mathfrak{P} \cap \mathfrak{o}_K = \mathfrak{p}$). In order to show that $\tilde{E}_p$ has a point of period p (i.e. $\tilde{E}_p$ is not supersingular), it suffices to do so for an elliptic curve isogenous to $\tilde{E}_p$. Changing E via isogeny as necessary, we may assume that $\theta : K \to \mathrm{End}(E)_{\mathbb{Q}}$ is a normalized embedding with $\theta(\mathfrak{o}_K) = \mathrm{End}(E)$. Choose $m \in \mathbb{Z}^+$ such that $\mathfrak{p}^m = \mu \mathfrak{o}_K$ and $\mathfrak{p}'^m = \mu' \mathfrak{o}_K$ are principal (note that $\mu\mu' = p^m$). We note that $\mu' \notin \mathfrak{p}$. For a differential form of the first kind ω, $\mu'\omega \not\equiv_{\mathfrak{P}} 0$, and since θ is normalized, $\widetilde{\theta(\mu')}_p$ is separable. Because the degree of $\theta(\mu')$ is a power of p, so is the degree of its reduction modulo $\mathfrak{P}$. We can thus conclude that $\tilde{E}_p$ has a non-trivial point of order p, and so it is not supersingular.

Let us now assume that $\mathrm{End}(E) \cong \mathfrak{o}$, where $\mathfrak{o}$ is an order in K with conductor $c = p^r c_0$, and $p \nmid c_0$. It is clear that the reduction map $\mathrm{End}(E) \to \mathrm{End}(\tilde{E}_p)$ is injective, so $\widetilde{\mathrm{End}(E)}_p \subset \mathrm{End}(\tilde{E}_p)$. By Theorem 3.1.8, $\mathrm{End}(\tilde{E}_p)_{\mathbb{Q}}$ is an imaginary quadratic field, so $\mathrm{End}(E)_{\mathbb{Q}} \overset{\cong}{\to} \mathrm{End}(\tilde{E}_p)_{\mathbb{Q}}$ is an isomorphism induced by induction.

Suppose that $p \nmid c$. For every prime $\ell \neq p$, we have an isomorphism $T_\ell(E) \overset{\cong}{\to} T_\ell(\tilde{E}_p)$, and by Lemma 3.1.14, $\mathrm{End}(E)$ and $\mathrm{End}(\tilde{E}_p)$ have the same localization at ℓ. Because $p \nmid c$,

$\mathfrak{o}_{(p)} = \mathfrak{o}_{K,(p)}$, so it is integrally closed and coincides with the localization of $\text{End}(\tilde{E}_p)$ at p. Because their localizations at each prime are the same, $\text{End}(E) \cong \text{End}(\tilde{E}_p)$.

$\square$

We shall conclude this section by stating the Deuring Lifting Theorem, which is historically significant, but which will not be used in this paper.

Theorem 3.1.18 (Deuring Lifting Theorem). *Let E_0 be an elliptic curve in characteristic p, with an endomorphism α_0 which is non-trivial. Then there exists an elliptic curve E defined over a number field, an endomorphism α of E, and a non-degenerate reduction of E at a place $\mathfrak{P}$ lying above p, such that E_0 is isomorphic to $\tilde{E}_p$, and α_0 corresponds to $\alpha(\mathfrak{P})$ under the isomorphism.*

3.2   The CM Method

The CM method is used to find elliptic curves of a given order. Since it is relatively easy to compute the order of an elliptic curve with CM in $\mathbb{Q}(\sqrt{-d})$, we can determine a value of d which produces an elliptic curve with a desired property (such as prime order), and then find an explicit formula for an elliptic curve with the desired number of points, as in [12].

In the CM method, first appearing in [1], we take a prime $p \geq 5$ and square-free positive integer $d \neq 1, 3, p$ and find integers a, b such that

$$4p = a^2 + db^2.$$

Then we output an elliptic curve $E/\mathbb{F}_p$ having CM in $\mathbb{Q}(\sqrt{-d})$ with order

$$\#E(\mathbb{F}_p) = p + 1 - a. \tag{3.1}$$

Note that a and b are unique up to choice of sign, so $\#E(\mathbb{F}_p) = p + 1 + a$ is also possible, and in fact, half of the quadratic twists of E have this order. The details of this method are of tangential interest to this paper, so we refer the interested reader to [1] or [12] for the details.

However, we will prove that (3.1) is valid, as it will be important in Section 4.2.

Theorem 3.2.1. *Let E be an elliptic curve defined over a number field, and let $\mathfrak{o}, p$, and $\mathfrak{P}$ be defined as in Theorem 3.1.17. If E has good reduction modulo $\mathfrak{P}$, then $\exists \pi \in \mathfrak{o}$ such that $p = \pi\bar{\pi}$ and*

$$\#\tilde{E}(\mathbb{F}_p) = p + 1 - (\pi + \bar{\pi}).$$

Proof. By Theorem 3.1.17, $\text{End}_{\mathbb{C}}(E) \cong \text{End}_{\overline{\mathbb{F}}_p}(\tilde{E})$, and the isomorphism (induced by reduction) preserves degrees. Then $\exists \pi \in \text{End}_{\mathbb{C}}(E)$ that corresponds to the Frobenius endomorphism $\pi_p \in \text{End}_{\overline{\mathbb{F}}_p}(\tilde{E})$ under reduction modulo p. Since degree is preserved,

$$\deg(\pi) = \deg(\pi_p) = p.$$

Over $\mathbb{C}$, we know that the degree of π is its norm, so $N(\pi) = p$. Therefore, $p = \pi\bar{\pi}$.

Recall from the proof of Theorem 2.3.1 that

$$\#\tilde{E}(\mathbb{F}_p) = \deg(1 - \pi_p).$$

Therefore,

$$\#\tilde{E}(\mathbb{F}_p) = \deg(1 - \pi) = N(1 - \pi) = (1 - \pi)(1 - \bar{\pi})$$
$$= \pi\bar{\pi} + 1 - \pi - \bar{\pi} = p + 1 - (\pi + \bar{\pi}),$$

as desired. $\qquad\square$

Note that

$$\mathfrak{o} = \begin{cases} \mathbb{Z}\left[\frac{1+\sqrt{-d}}{2}\right] & \text{if } d \equiv_4 1 \\ \mathbb{Z}[\sqrt{-d}] & \text{otherwise.} \end{cases}$$

Because π is in the ring of integers of $\mathbb{Q}(\sqrt{-d})$ (recall that $d \neq 1, 3$), (3.1) follows immediately.

3.3   Special Cases: $j = 0$ and $j = 1728$

We noted in Section 3.2 that (3.1) does not apply when $d = 1, 3$. In fact, it applies in those cases, except when $j(E) = 0, 1728$. For the other three cases (see Table 2.1), we can use (3.1) to compute the order. When $j(E) = 1728$, $\tilde{E}$ can take one of four orders, and when $j(E) = 0$, it can take one of six orders, which can be computed using classical methods.

For $j(E) = 1728$, changing variables as necessary, we can write $E : y^2 = x^3 - kx$ for some integer k. Then $(x, y) \mapsto (-x, iy)$ is an endomorphism (in fact, an isomorphism), and it is contained in the ring of integers of $\mathbb{Q}(i) = \mathbb{Q}(\sqrt{-1})$, so $d = 1$. It should be noted that this endomorphism does not constitute multiplication by a real integer because it is of order 4 - so it is a fourth root of unity (i.e. $\pm i$).

Theorem 3.3.1. *Let $E : y^2 = x^3 - kx$ be an elliptic curve. If $p \geq 5, p \equiv_4 3$, then $\#\tilde{E}(\mathbb{F}_p) = p+1$. Otherwise ($p \equiv_4 1$), find integers a, b such that $p = a^2 + b^2$ with b even and $a + b \equiv_4 1$. Then*

$$\#\tilde{E}(\mathbb{F}_p) = \begin{cases} p + 1 - 2a & \text{if } \left(\frac{k}{p}\right)_4 = 1, \\ p + 1 + 2a & \text{if } \left(\frac{k}{p}\right)_4 = -1, \\ p + 1 \pm 2b & \text{otherwise,} \end{cases}$$

where $\left(\frac{\cdot}{\cdot}\right)_4$ is the quartic residue symbol.

Proof. We leave the proof of this result to [18], as it will not be used later in this paper (because the order is only prime when $\#\tilde{E}(\mathbb{F}_5) = 2$). $\qquad\square$

For $j(E) = 0$, changing variables as necessary, we can write $E : y^2 = x^3 + k$ for some integer k.

In order to determine the orders, we will need two lemmas. We define a **multiplicative character** χ by choosing a primitive root g modulo p. Then for all $k | (p - 1)$, we set $\chi_k(g^j) = e^{2ij\pi/k}$.

Lemma 3.3.2. *Let $p \equiv_3 1$ be prime and let $x \in \mathbb{F}_p^\times$. Then*

$$\#\{u \in \mathbb{F}_p^\times | u^t = x\} = \sum_{\ell=0}^{t-1} \chi_t(x)^\ell$$

for $t \in \{2, 3, 6\}$.

Proof. Since $p \equiv_3 1$, we also have that $p \equiv_6 1$, and so there are 6 sixth roots of 1 in $\mathbb{F}_p^\times$. Therefore, if there is a solution to $u^6 \equiv x$, there are 6 solutions. Write $x \equiv g^j \mod p$. Then x is a sixth power modulo p if and only if $j \equiv_6 0$. We have

$$\sum_{\ell=0}^{5} \chi_6(x)^\ell = \sum_{\ell=0}^{5} e^{ij\ell\pi/3},$$

which evaluates to 6 if $j \equiv_6 0$, and to 0 otherwise, giving us exactly the number of u for which $u^6 \equiv x$. This proves the case for $t = 6$; the proofs for $t = 2$ and $t = 3$ follow similarly. $\square$

Note that Lemma 3.3.2 holds true for all $t | (p - 1)$, although we only need the cases given above. In general, we have to consider $\chi_{\gcd(p-1,t)}$ to determine the number of roots for arbitrary t.

Lemma 3.3.3. *Let $p \equiv_3 1$ be prime. Then*

$$\sum_{b \in \mathbb{F}_p^\times} \chi_6(b)^\ell = \begin{cases} p - 1 & \text{if } \ell \equiv_6 0, \\ 0 & \text{otherwise.} \end{cases}$$

Proof. If $\ell \equiv_6 0$, all the terms in the sum are 1, so the sum is $p - 1$. If $\ell \not\equiv_6 0$, then $\chi_6(g)^\ell \neq 1$. Multiplying by g permutes the elements of $\mathbb{F}_p^\times$, so

$$\chi_6(g)^\ell \sum_{b \in \mathbb{F}_p^\times} \chi_6(b)^\ell = \sum_{b \in \mathbb{F}_p^\times} \chi_6(gb)^\ell = \sum_{c \in \mathbb{F}_p^\times} \chi_6(c)^\ell,$$

which is the original sum. As $\chi_6(g)^\ell \neq 1$, the sum must be 0. $\square$

Theorem 3.3.4. *Let $p > 3$ be an odd prime and let $k \not\equiv_p 0$. Let E be the elliptic curve $y^2 = x^3 + k$.*

1. *If $p \equiv_3 2$, then $\#\tilde{E}(\mathbb{F}_p) = p + 1$.*

28

2. If $p \equiv_3 1$, write $p = a^2 + 3b^2$,[1] where a, b are integers with b positive and $a \equiv_3 -1$. Then

$$\#\tilde{E}(\mathbb{F}_p) = \begin{cases} p + 1 + 2a & \text{if } k \text{ is a sixth power mod } p \\ p + 1 - 2a & \text{if } k \text{ is a CR, but not a QR, mod } p \\ p + 1 - a \pm 3b & \text{if } k \text{ is a QR, but not a CR, mod } p \\ p + 1 + a \pm 3b & \text{if } k \text{ is neither a QR nor a CR mod } p. \end{cases}$$

Here, QR and CR denote quadratic residue and cubic residue, respectively.

Proof. The case where $p \equiv_3 2$ is easy. Since every value modulo p is a cubic residue (since 3 is coprime to $p - 1$), then over all $x \in \mathbb{F}_p$, $f(x) = x^3 + k$ takes on every value modulo p. Of these, $f(x) = 0$ is a special case, corresponding to the point $(x, 0)$, and the other values of $f(x)$ are split evenly into groups: quadratic residues (QR), and quadratic non-residues (QNR). Each QR corresponds to two points on E, whereas each QNR corresponds to zero points on E. Summing, we get p points. Adding the point at infinity $\mathcal{O}$, this gives us

$$\#\tilde{E}(\mathbb{F}_p) = p + 1.$$

Now we consider $p \equiv_3 1$. For this case, we choose a primitive root g modulo p to define $\chi_6(g^j) = e^{ij\pi/3}$. Then $\chi_6^2 = \chi_3$ and $\chi_6^3 = \chi_2$.

We now show that the number of points on E can be expressed in terms of Jacobi sums. By separating out the terms where $x = 0$ and $y = 0$, we find that the number of points is

$$\#\{\mathcal{O}\} + \#\{y | y^2 = k\} + \#\{x | x^3 = -k\} + \sum_{a+b=k; a,b \neq 0} \#\{y | y^2 = a\} \#\{x | x^3 = -b\}$$

By Lemma 3.3.2, the first three terms are 1, $\sum_{j=0}^{1} \chi_2(k)^j$, and $\sum_{l=0}^{2} \chi_3(-k)^l$, respectively. The latter summation expands to

$$\sum_{a \neq 0, k} \sum_{j=0}^{1} \chi_2(a)^j \sum_{l=0}^{2} \chi_3(a - k)^l$$

$$= \sum_{a \neq 0, k} [\chi_2(a)^0 + \chi_2(a)^1][\chi_3(a - k)^0 + \chi_3(a - k)^1 + \chi_3(a - k)^2]$$

$$= \sum_{a \neq 0, k} [1 + \chi_2(a) + \chi_3(a - k) + \chi_2(a)\chi_3(a - k) + \chi_3(a - k)^2 + \chi_2(a)\chi_3(a - k)^2]$$

$$= (p - 2) - \chi_2(k) - \chi_3(-k) + \sum_{a \neq 0, k} \chi_2(a)\chi_3(a - k) - \chi_3(-k)^2 + \sum_{a \neq 0, k} \chi_2(a)\chi_3(a - k)^2.$$

---

[1]It is common to formulate this theorem using $p = a^2 - ab + b^2$ instead, although the notational choice given in Thoerem 3.3.4 is more consistent with the CM method.

We have used Lemma 3.3.3 to replace the sums of single characters χ with the negative of the value $\chi(\pm k)$ that was omitted from the sum. Combining with the terms from before, this simplifies to:

$$p + 1 + \sum_{a \neq 0, k} \chi_2(a)\chi_3(a - k) + \sum_{a \neq 0, k} \chi_2(a)\chi_3(a - k)^2$$

$$= p + 1 + \chi_6(-1)^2 \sum_{a \neq 0, k} \chi_2(a)\chi_3(k - a) + \chi_6(-1)^4 \sum_{a \neq 0, k} \chi_2(a)\chi_3(k - a)^2$$

$$= p + 1 + \chi_6(-1)^2 \chi_6(k)^{-1} \sum_{a/k \neq 0, 1} \chi_2(a/k)\chi_3((k - a)/k)$$

$$+ \chi_6(-1)^4 \chi_6(k)^{-1} \sum_{a/k \neq 0, 1} \chi_2(a/k)\chi_3((k - a)/k)^2$$

$$= p + 1 + \chi_6(k)^{-1} J(\chi_2, \chi_3) + \chi_6(k)^{-1} J(\chi_2, \chi_3^2),$$

where $J(\chi_2, \chi_3)$ is the Jacobi character, and the last equality holds because $\chi_6(-1)^2 = \chi_3(-1)$. $-1 = (-1)^3$, so $\chi_6(-1)^2 = \chi_6(-1)^4 = 1$.

We note that if we write

$$\#E(\mathbb{F}_p) = p + 1 - \alpha - \bar{\alpha},$$

then $\alpha = -\chi_6(k)^{-1} J(\chi_2, \chi_3)$. In [3], a_3 and b_3 are defined such that $a_3^2 + 3b_3^2 = p$ and $a_3 \equiv -1$ mod 3 (p. 103). Then, in Table 3.1.2 (p. 107), it is stated that $J(\chi_2, \chi_3) = a_3 + ib_3\sqrt{3}$.

If k is a sextic residue, then $\alpha = -a_3 - ib_3\sqrt{3}$, so $\#\tilde{E}(\mathbb{F}_p) = p + 1 + 2a_3$. If k is a cubic residue, but not a quadratic residue, then $\#\tilde{E}(\mathbb{F}_p) = p + 1 - 2a_3$. If k is a quadratic residue, but not a cubic residue, then $\alpha = \frac{-1 \pm i\sqrt{3}}{2} \cdot (a_3 + ib_3\sqrt{3}) = \frac{-a_3 \mp 3b_3}{2} + \frac{-b_3\sqrt{3} \pm a_3\sqrt{3}}{2}i$. Thus, $\#\tilde{E}(\mathbb{F}_p) = p + 1 - a_3 \pm 3b_3$. The final case, where k is neither a quadratic residue nor a cubic residue, follows the same way, and we find that $\#\tilde{E}(\mathbb{F}_p) = p + 1 + a_3 \mp 3b_3$.

The end result is that $\#\tilde{E}(\mathbb{F}_p) = p + 1 - \pi - \bar{\pi}$, where $\pi = -\chi_6(k)^{-1} J(\chi_2, \chi_3)$. This Jacobi sum is evaluated as $J(\chi_2, \chi_3) = a + ib\sqrt{3}$, with $a \equiv_3 -1$ [3]. $\square$

Chapter 4

Elliptic Reciprocity

We will use Größencharakter to prove our main results in Section 4.2. Section 4.1 will be devoted to the introduction of Größencharakter (also known as Hecke characters), including the proofs of several results which will be of use in Section 4.2. Section 4.2 closely follows [2].

4.1 Größencharakter

Let K be a number field, and define K_ν to be the completion of K at the absolute value ν. Then we let R_ν be the ring of integers of K_ν for non-archimedean ν, and $R_\nu = K_\nu$ otherwise. We define the **idèle group** of K to be

$$\mathbf{A}_K^* = \prod_\nu{}' K_\nu^*,$$

where the product is over all possible absolute values ν and the prime means that the product is restricted relative to the R_ν's. Thus, the ν-component of an idèle s, s_ν, is an element of R_ν, for all but finitely many ν.

The ideal of s is a fractional ideal of K:

$$(s) = \prod_{\mathfrak{p}} \mathfrak{p}^{\operatorname{ord}_\mathfrak{p} s_\mathfrak{p}},$$

where $\mathfrak{p}$ denotes a prime ideal. When $K = \mathbb{Q}$, we can define $N_s \in \mathbb{Q}^*$ such that $(s) = s\mathbb{Z} = N_s\mathbb{Z}$ and $\operatorname{sign}(N_s) = \operatorname{sign}(s_\infty)$, where s_∞ is the value of s at the archimedean component. The cyclotomic character

$$\chi : \operatorname{Gal}(K^{\mathrm{ab}}/K) \longrightarrow \hat{\mathfrak{o}}_K^\times \cong \mathbf{A}_K^*,$$

where K^{ab} is the maximal abelian extension of K, is defined as $\sigma(\zeta) =: \zeta^{\chi(\sigma)}$ [6]. From its inverse, we can define the continuous homomorphism

$$\mathbf{A}_K^* \longrightarrow \operatorname{Gal}(K^{\mathrm{ab}}/K), \qquad s \mapsto [s, K].$$

If (s) is not divisible by any primes ramifying in L, an extension field of K, then $[s, K]|_L = ((s), L/K)$ is the Artin map of (s) and we call $[\cdot, K]$ the **reciprocity map** for K [14].

We now state the Main Theorem of Complex Multiplication:

Theorem 4.1.1. *Fix a quadratic imaginary field K over $\mathbb{Q}$ with ring of integers $\mathfrak{o}_K$, an elliptic curve $E/\mathbb{C}$ with $\mathrm{End}(E) \cong \mathfrak{o}_K$, $\sigma \in \mathrm{Aut}(\mathbb{C})$, and $s \in \mathbf{A}_K^*$ such that $[s, K] = \sigma|_{K^{ab}}$. Let $\mathfrak{a}$ be a fractional ideal of K, and fix a complex analytic isomorphism*

$$f : \mathbb{C}/\mathfrak{a} \longrightarrow E(\mathbb{C}).$$

Then there exists a unique complex analytic isomorphism

$$f' : \mathbb{C}/s^{-1}\mathfrak{a} \longrightarrow E^{\sigma}(\mathbb{C})$$

such that the following diagram commutes:

$$
\begin{CD}
K/\mathfrak{a} @>{s^{-1}}>> K/s^{-1}\mathfrak{a} \\
@V{f}VV @VV{f'}V \\
E(\mathbb{C}) @>{\sigma}>> E^{\sigma}(\mathbb{C}).
\end{CD}
$$

The proof is long and unenlightening, so we omit it here, but the interested reader may find it in Section II.8 of [14].

We define a Größencharakter

$$\psi : \mathbf{A}_L^* \longrightarrow \mathbb{C}^*$$

to be a continuous homomorphism such that $\psi(L^*) = 1$. If $L = \mathbb{Q}$, $\psi(s) = N_s s_\infty^{-1}$ is a Größencharakter [14]. We will use Theorem 4.1.1 to define a Größencharakter ψ which is associated to an elliptic curve E with CM.

Theorem 4.1.2. *Let E/L be an elliptic curve with CM by $\mathfrak{o}_K$ for $K = \mathbb{Q}(\sqrt{-d}) \subset L$. Let $x \in \mathbf{A}_L^*$ and $s = N_K^L x \in \mathbf{A}_K^*$. Then $\exists! \alpha = \alpha_{E/L}(x) \in K^*$ such that*

(i) *$\alpha \mathfrak{o}_K = (s)$, the ideal of s,*

(ii) *for any fractional ideal $\mathfrak{a} \subset K$ and any analytic isomorphism*

$$f : \mathbb{C}/\mathfrak{a} \to E(\mathbb{C}),$$

the following diagram commutes:

$$K/\mathfrak{a} \xrightarrow{\alpha s^{-1}} K/\mathfrak{a}$$

$$\downarrow f \qquad\qquad \downarrow f$$

$$E(L^{\mathrm{ab}}) \xrightarrow{[x,L]} E(L^{\mathrm{ab}}).$$

Proof. Let $L' = L(E_{\mathrm{tors}})$, so that $K^{\mathrm{ab}} \subset L' \subset L^{\mathrm{ab}}$ [14]. Choose $\sigma \in \mathrm{Aut}(\mathbb{C})$ such that

$$\sigma\Big|_{L^{\mathrm{ab}}} = [x, L]. \tag{4.1}$$

By Artin reciprocity, $\sigma\Big|_{K^{\mathrm{ab}}} = [s, K]$. By 4.1.1, we find analytic isomorphism f' such that

$$K/\mathfrak{a} \xrightarrow{s^{-1}} K/s^{-1}\mathfrak{a}$$

$$\downarrow f \qquad\qquad \downarrow f'$$

$$E(\mathbb{C}) \xrightarrow{\sigma} E^{\sigma}(\mathbb{C})$$

is a commutative diagram. Since σ fixes L, $E^{\sigma} = E$, so $\mathfrak{a}$ and $s^{-1}\mathfrak{a}$ are homothetic. Therefore, $\exists \beta \in K^*$ such that $\beta s^{-1}\mathfrak{a} = \mathfrak{a}$ and our commutative diagram becomes

$$K/\mathfrak{a} \xrightarrow{\beta s^{-1}} K/\mathfrak{a}$$

$$\downarrow f \qquad\qquad \downarrow f''$$

$$E(\mathbb{C}) \xrightarrow{\sigma} E(\mathbb{C}).$$

We note that $[\xi] = f'' \circ f^{-1} \in \mathrm{Aut}(E)$, and we set $\alpha = \xi\beta$. Now (ii) follows immediately from 4.1. Furthermore,

$$\alpha s^{-1}\mathfrak{a} = \beta s^{-1}\mathfrak{a} = \mathfrak{a},$$

so $\alpha \mathfrak{o}_K = (s)$, proving (i).

It remains to show that α is unique and independent of choice of f. Suppose α' also satisfies (i) and (ii). Because f and $[x, L]$ are isomorphisms,

$$K/\mathfrak{a}$$
$$\alpha s^{-1} \swarrow \qquad \searrow \alpha' s^{-1}$$
$$K/\mathfrak{a} =\!=\!=\!=\!= K/\mathfrak{a}$$

is a commutative triangle. From this we see that $[\alpha'\alpha^{-1}]$ is the identity map on $K/\mathfrak{a}$, so $\alpha' = \alpha$, and α is unique. Now, suppose that $f' : \mathbb{C}/\mathfrak{a}' \to E(\mathbb{C})$ is an analytic isomorphism with $\mathfrak{a} = \gamma\mathfrak{a}'$ for some $\gamma \in K^*$. Then $[\xi] = f' \circ f^{-1} \in \operatorname{Aut}(E)$, so $f'(z) = f(\xi\gamma z)$, and (ii) implies that $\forall t \in K/\mathfrak{a}$,

$$f'(t)^{[x,L]} = f(\xi\gamma t)^{[x,L]} = f(\alpha s^{-1}\xi\gamma t) = f'(\alpha s^{-1}t).$$

Hence (ii) is true for f', and α does not depend on f. $\square$

Theorem 4.1.2 tells us that the map

$$\alpha_{E/L} : \mathbf{A}_L^* \to K^* \subset \mathbb{C}^*$$

is a homomorphism, and furthermore, we can compute $\alpha_{E/L}(x_\beta) = N_K^L\beta$, where $\beta \in L^*$ and x_β is the corresponding idèle [14]. This means that $\alpha_{E/L}$ cannot be a Größencharakter (because $\alpha \neq 1$ on L^*), although we can make it one via a slight modification.

Theorem 4.1.3. *Assume the same hypotheses as in Theorem 4.1.2, and for any $s \in \mathbf{A}_K^*$, let $s_\infty \in \mathbb{C}^*$ be the component of s corresponding to the unique archimedean absolute value on K. Then*

(a) *The map*
$$\psi_{E/L} : \mathbf{A}_L^* \to \mathbb{C}^*, \psi_{E/L}(x) = \alpha_{E/L}(x)N_K^L(x^{-1})_\infty$$
   is a Größencharakter of L.

(b) *Let $\mathfrak{P}$ be a prime of L. Then $\psi_{E/L}$ is unramified ($\psi(\mathfrak{o}_\mathfrak{P}^*) = 1$) at $\mathfrak{P}$ if and only if E has good reduction at $\mathfrak{P}$.*

Proof. See Theorem II.9.2 of [14]. $\square$

We can use ψ to compute the order of an elliptic curve over a finite field.

Lemma 4.1.4. *Assume the same hypotheses as in Theorems 4.1.2 and 4.1.3, and assume that E has good reduction at $\mathfrak{P}$. Let $\tilde{E}$ be the reduction of E modulo $\mathfrak{P}$, and let $\pi_\mathfrak{P}$ be the $N_\mathbb{Q}^L\mathfrak{P}$-power Frobenius endomorphism. Then*

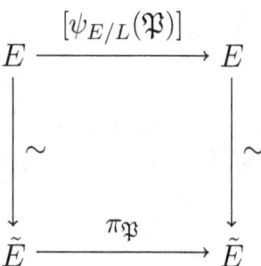

commutes, where $\sim$ denotes reduction modulo $\mathfrak{P}$.

Proof. Let $x \in \mathbf{A}_L^*$ such that $\psi_{E/L}(\mathfrak{P}) = \psi_{E/L}(x) = \alpha_{E/L}(x)$. By Theorem 4.1.2ii, $\forall t \in K/\mathfrak{a}$,

$$f(t)^{[x,L]} = [\psi_{E/L}(x)]f(N_K^L x^{-1} t).$$

Fix $m \in \mathbb{Z}$ such that $\mathfrak{P} \nmid m$. It can be shown that $\forall t \in m^{-1}\mathfrak{a}/\mathfrak{a}$, $N_K^L x^{-1} t = t$ [14] Theorem II.9.3. Therefore,

$$f(t)^{[x,L]} = [\psi_{E/L}(\mathfrak{P})]f(t)$$

and because $[x, L]$ reduces to the $q_{\mathfrak{P}}$-power Frobenius map,

$$\pi_{\mathfrak{P}}(\widetilde{f(t)}) = \widetilde{f(t)^{[x,L]}} = [\widetilde{\psi_{E/L}(\mathfrak{P})}]\widetilde{f(t)}.$$

Since our choice of m was arbitrary, and we can define an endomorphism of $\tilde{E}$ by specifying its value on the torsion points, we determine that $\pi_{\mathfrak{P}} = [\widetilde{\psi_{E/L}(x)}]$, from which the fact that the diagram commutes is immediate. $\qquad\square$

Theorem 4.1.5. *With notation as in Lemma 4.1.4,*

(a) $N_{\mathbb{Q}}^L \mathfrak{P} = N_{\mathbb{Q}}^K(\psi_{E/L}(\mathfrak{P}))$,

(b) $\#\tilde{E}(\mathbb{F}_{\mathfrak{P}}) = N_{\mathbb{Q}}^L \mathfrak{P} + 1 - \psi_{E/L}(\mathfrak{P}) - \overline{\psi_{E/L}(\mathfrak{P})}$,

(c) $a_{\mathfrak{P}} = \psi_{E/L}(\mathfrak{P}) + \overline{\psi_{E/L}(\mathfrak{P})}$.

Proof. For (a), we compute

$$N_{\mathbb{Q}}^L \mathfrak{P} = \deg \pi_{\mathfrak{P}} = \deg[\widetilde{\psi_{E/L}(\mathfrak{P})}] = \deg[\psi_{E/L}(\mathfrak{P})] = N_{\mathbb{Q}}^K(\psi_{E/L}(\mathfrak{P})).$$

For (b),

$$\begin{aligned}
\#\tilde{E}(\mathbb{F}_{\mathfrak{P}}) &= \#\ker(1 - \pi_{\mathfrak{P}}) = \deg(1 - \pi_{\mathfrak{P}}) \\
&= \deg[1 - \widetilde{\psi_{E/L}(\mathfrak{P})}] = \deg[1 - \psi_{E/L}(\mathfrak{P})] \\
&= N_{\mathbb{Q}}^K(1 - \psi_{E/L}(\mathfrak{P})) = (1 - \psi_{E/L}(\mathfrak{P}))(1 - \overline{\psi_{E/L}(\mathfrak{P})}) \\
&= 1 - \psi_{E/L}(\mathfrak{P}) - \overline{\psi_{E/L}(\mathfrak{P})} + N_{\mathbb{Q}}^L \mathfrak{P}.
\end{aligned}$$

Part (c) is an immediate consequence of (b) and the definition of $a_{\mathfrak{P}}$. $\qquad\square$

We now focus in on the case that $L = \mathbb{Q}$. Note that $a_{\mathfrak{P}}$ is the trace of the Frobenius endomorphism, and we can also use $\mathrm{Tr}(\psi_{E/L}(\mathfrak{P}))$ to represent the same quantity [15]. Recall from Section 3.1 that if p splits in K, then we can write $p\mathfrak{o}_K = \mathfrak{p}\bar{\mathfrak{p}}$, so $\mathbb{F}_p = \mathbb{F}_{\mathfrak{p}}$ and

$$q = \#\tilde{E}_p(\mathbb{F}_p) = \#\tilde{E}_{\mathfrak{p}}(\mathbb{F}_{\mathfrak{p}}) = N_{\mathbb{Q}}^K(1 - \psi_{E/L}(\mathfrak{P})). \tag{4.2}$$

This tells us that q also splits in K ($q\mathfrak{o}_K = \mathfrak{q}\bar{\mathfrak{q}}$), so

$$q = N_{\mathbb{Q}}^K(\psi_{E/L}(\mathfrak{Q})). \tag{4.3}$$

From Equations (4.2) and (4.3), we immediately see that

$$\psi_{E/L}(\mathfrak{Q}) = \begin{cases} u(1 - \psi_{E/L}(\mathfrak{P})) & \text{or} \\ \overline{u(1 - \psi_{E/L}(\mathfrak{P}))} & \text{for some } u \in \mathfrak{o}_K^*. \end{cases} \tag{4.4}$$

For $K \neq \mathbb{Q}(\sqrt{-1}), \mathbb{Q}(\sqrt{-3})$, $\mathfrak{o}_K^* = \{\pm 1\}$, so

$$\mathrm{Tr}(\psi_{E/L}(\mathfrak{Q})) = \pm\mathrm{Tr}(1 - \psi_{E/L}(\mathfrak{P})) \tag{4.5}$$
$$= \pm(2 - \mathrm{Tr}(\psi_{E/L}(\mathfrak{P}))) \qquad \text{(by linearity)} \tag{4.6}$$
$$= \pm(2 - (p + 1 - q)) = \pm(q + 1 - p). \tag{4.7}$$

From this we immediately notice the following convenient result, which is Theorem 6.1 of [15]. Recall that if p does not split in $\mathfrak{o}_K$, then E is supersingular, and in particular, its order over $\mathbb{F}_p$ is $p + 1$.

Theorem 4.1.6. *Let E/L have CM, and let $j(E) \neq 0$. If $\#\tilde{E}(\mathbb{F}_p) = q$ is prime for some prime $p \geq 5$ of good reduction, and if $q \geq 5$ is also of good reduction for E, then*

$$\#\tilde{E}(\mathbb{F}_q) = \begin{cases} p & \text{or} \\ 2q + 2 - p. \end{cases}$$

This is a generalization of Theorem 6.1 of [15], which only considers the case that E is defined over $\mathbb{Q}$ (although nothing in the proof requires this restriction), which would limit us to the curves (up to isomorphism/twist) in Table 2.1. By generalizing to defining E over L, we allow fields K with class number other than one.

Proof. Note that by Theorem 3.3.1, if E has CM in $\mathbb{Q}(\sqrt{-1})$, then its order is even, so it is not prime. We also note that if $j(E) = -12288000$, E has CM in $\mathbb{Q}(\sqrt{-3})$, but Equation (4.7) still holds [15]. By Equation (4.7) and Theorem 4.1.5,

$$\#\tilde{E}(\mathbb{F}_q) = q + 1 - \mathrm{Tr}(\psi_{E/L}(\mathfrak{Q})) = q + 1 \mp (q + 1 - p).$$

The theorem follows immediately. $\qquad\qquad\qquad\qquad\qquad\qquad\qquad\qquad\square$

For $j(E) = 0$, we have a similar result, although it is more complicated because $\mathfrak{o}_K^*$ is the set of sixth roots of unity in this case. We have that for $E : y^2 = x^3 + k$, its Größencharakter $\psi_{E/L}(\mathfrak{P}) = -\left(\frac{4k}{\mathfrak{p}}\right)^{-1}\pi$, where π is the generator from Corollary 3.1.9 [14], [15]. Since π is primary, $\pi \equiv_{3\mathfrak{o}_K} 2$.

Corollary 7.6 of [15] will give us an analogous result to Theorem 4.1.6 for curves with $j(E) = 0$:

Theorem 4.1.7. *Let E/L be an elliptic curve with $j(E) = 0$, and let $p, q \geq 5$ be primes of good reduction for E. If $\#\tilde{E}(\mathbb{F}_p) = q$, then:*

(a) *there exists $A \in \mathbb{Z}$ such that*

$$A^2 = \frac{2pq + 2p + 2q - p^2 - q^2 - 1}{3},$$

(b) *the trace $a_q(E) = q + 1 - \#\tilde{E}(\mathbb{F}_q)$ is one of six values:*

$$a_q(E) = \begin{cases} \pm(q + 1 - p) \\ \frac{\pm(q+1-p)\pm 3A}{2} \end{cases}.$$

Proof. We know from the proof of Theorem 3.3.4 that $\mathfrak{p} = \psi_{E/L}(\mathfrak{P}) \in \mathfrak{o}_K = \mathbb{Z}\left[\frac{1+\sqrt{-3}}{2}\right]$, and since $\text{Tr}(\psi_{E/L}(\mathfrak{P})) = a_p(E)$, so $\exists A \in \mathbb{Z}$ such that $\psi_{E/L}(\mathfrak{P}) = \frac{a_p(E)+A\sqrt{-3}}{2}$. Since $p = N_{\mathbb{Q}}^K(\mathfrak{P}) = \frac{a_p(E)^2+3A^2}{4}$, we find that $A^2 = \frac{4p-(p+1-q)^2}{3}$, which proves part (a).

For (b), we see from the proof of Theorem 3.3.4 and from Equation (4.4) that

$$\text{Tr}(\psi_{E/L}(\mathfrak{Q})) = \text{Tr}(\zeta(1 - \psi_{E/L}(\mathfrak{P}))),$$

where ζ is a sixth root of unity. Using the value of $\psi_{E/L}(\mathfrak{P})$ above, we get the six values listed for (b). $\qquad\square$

4.2 Aliquot Cycles for Elliptic Curves with Complex Multiplication

4.2.1 Elliptic Pairs

As always, all prime numbers p and q are without further comment assumed to be at least 5. We let L be an extension field of $\mathbb{Q}$, and for a square-free positive integer d, we let $K = \mathbb{Q}(\sqrt{-d})$ and $\mathfrak{o}_K$ be the ring of integers in K.

Definition 4.2.1. *For a square-free positive integer d, define an **elliptic pair over** d to be an ordered pair (p, q) of prime numbers such that there exists an elliptic curve E/L with complex multiplication (CM) by $\mathfrak{o}_K$ having order q when considered over $\mathbb{F}_p$. We write $(p, q)_d$ to denote the fact that the ordered pair (p, q) is an elliptic pair over d. The witnessing curve E is referred to as a **representative curve** of the pair.*

For example, $(7, 13)_3$ is an elliptic pair because the representative curve $E : y^2 = x^3 + 3$ has order 13 when considered over the field $\mathbb{F}_7$.

As stated, Definition 4.2.1 is not symmetric with respect to p and q. We will see in Theorem 4.2.7 below that we may treat it as symmetric, however.

Definition 4.2.2. *An **elliptic prime (over** d**)** is a prime number which is the first entry in an elliptic pair over d.*

In Definition 4.2.1, p is an elliptic prime over d. In the example following Definition 4.2.1, 7 is an elliptic prime over 3.

Lemma 4.2.3. *Let E be an elliptic curve CM by $\mathfrak{o}_K$. If $d \not\equiv_8 3$, then $\#\tilde{E}(\mathbb{F}_p)$ is composite.*

In other words, if p is an elliptic prime over d, then $d \equiv_8 3$.

Proof. If $d \not\equiv_4 3$, then any representative curve E/L trivially has two-torsion by the CM method ($\#\tilde{E}(\mathbb{F}_p) = p + 1 \pm 2a$, where $p = a^2 + db^2$), so p cannot be an elliptic prime over d.

Let $d \equiv_4 3$, and let p be a prime. We may assume $d > 3$. Assuming that p is not prime in $\mathfrak{o}_K$, let a and b be the unique positive integers such that $4p = a^2 + db^2$.

Let E be a representative curve witnessing that p is an elliptic prime over d. If E has CM by $\mathfrak{o}_K$, then $\#\tilde{E}(\mathbb{F}_p) = p + 1 \pm a$. Then, as $p + 1$ is even and $\#\tilde{E}(\mathbb{F}_p)$ is a prime number, a must be odd. If it were the case that $d \equiv_8 7$, then $4 \equiv_8 4p = a^2 + db^2 \equiv_8 a^2 - b^2$. But since a is odd, $a^2 \equiv_8 1$, implying $b^2 \equiv_8 5$. But the congruence $x^2 \equiv_8 5$ does not have a solution, a contradiction. Thus, as $d \equiv_4 3$ we must have $d \equiv_8 3$. $\qquad\square$

We must treat the case of $d = 3$ separately. The curves E with CM in $K = \mathbb{Q}(\sqrt{-3})$ are precisely those with $j(E) \in \{0, 54000, -12288000\}$ [14]. Curves with $j(E) = 54000$ trivially have 2-torsion, but some curves with $j(E) = 0, -12288000$ have trivial torsion (over $\mathbb{Q}$), so we must consider those. Curves with $j = -12288000$ are covered fully in [15], and they can be treated in exactly the same way as other curves with CM, so we omit a full discussion of them here. When $j(E) = 0$, E has the form $E : y^2 = x^3 + k$ for $k \in \mathbb{Z}$, $\gcd(k, p) = 1$. Since these cases exhibit strikingly different behavior, for the rest of the paper we will write $d = *$ to specify that $d = 3$ and $j = 0$. Now, $p = a^2 + 3b^2$ and $\#\tilde{E}(\mathbb{F}_p)$ takes on one of six values:

Theorem 4.2.4. *Let $p > 3$ be an odd prime and let $k \not\equiv_p 0$. Let E be the elliptic curve $y^2 = x^3 + k$*

(a) *If $p \equiv_3 2$, then $\#\tilde{E}(\mathbb{F}_p) = p + 1$.*

(b) *If $p \equiv_3 1$, write $p = a^2 + 3b^2$, where a, b are integers with b positive and $a \equiv_3 -1$. Then*

$$\#\tilde{E}(\mathbb{F}_p) = \begin{cases} p + 1 + 2a & \text{if } k \text{ is a sixth power mod } p \\ p + 1 - 2a & \text{if } k \text{ is a CR, but not a QR, mod } p \\ p + 1 - a \pm 3b & \text{if } k \text{ is a QR, but not a CR, mod } p \\ p + 1 + a \pm 3b & \text{if } k \text{ is neither a QR nor a CR mod } p. \end{cases}$$

Here, QR and CR denote quadratic residue and cubic residue, respectively.

Proof. This is a restatement of Theorem 3.3.4, and the proof can be found there. $\qquad\square$

Corollary 4.2.5. *Let $p \equiv_3 1$ be a prime and let $E : y^2 = x^3 + k$ be an elliptic curve with $k \not\equiv_p 0$. If $\#\tilde{E}(\mathbb{F}_p)$ is prime, then k is neither a QR nor CR, except in the case where $p = 7$ and $k \equiv_7 4$ (in which case $\#\tilde{E}(\mathbb{F}_7) = 3$).*

Corollary 4.2.6. *Let $p > 7$ be a prime, and let $E : y^2 = x^3 + k$ be an elliptic curve. Then $\#\tilde{E}(\mathbb{F}_p)$ can take on at most 2 prime values (namely $p + 1 + a \pm 3b$ from Theorem 4.2.4), both of which are congruent to 1 modulo 3.*

We note that p can be in an elliptic pair with at most two primes given d. Note that the primes for $d = 3$ (for $j(E) \neq 0$) are a subset of those for $d = *$ because if $j(E) \neq 0$, at most one of the two possible orders can be prime (see Theorem 4.2.21).

Theorem 4.2.7 (The Law of Elliptic Reciprocity). *Let E have CM by $\mathfrak{o}_K$, and let p, q be primes such that $j(E)$ is defined both modulo p and modulo q. If $(p, q)_d$ is an elliptic pair, then so too is $(q, p)_d$.*

Proof. This follows immediately from Theorems 4.1.6 and 4.1.7. Let E be a representative curve of $(p, q)_d$. We find an appropriate quadratic (or sextic if $d = *$) twist of E modulo q, and then we use the Chinese remainder theorem to set the coefficients.

Note that (p, q) is an amicable pair (see [15] for definition, as this fact will not be used later in this paper) for the chosen twist of E. $\qquad\square$

One consequence of Theorem 4.2.7 is the following:

Theorem 4.2.8. *Let $(p, q)_d$ be an elliptic pair with $p < q$ and $d \neq *$, and a_p, b_p, a_q, b_q be the unique positive integers such that $4p = a_p^2 + db_p^2$ and $4q = a_q^2 + db_q^2$. Then $a_q = a_p + 2$ and $b_q = b_p$.*

Proof. We know that $q = p + 1 + a_p$ and by Theorem 4.2.7 that $p = q + 1 - a_q$, so $a_q = a_p + 2$. Then because $\frac{1}{4}(a_p^2 + db_p^2) = p$ and $\frac{1}{4}(a_q^2 + db_q^2) = q$,

$$\frac{a_p^2 + db_p^2}{4} + 1 + a_p = \frac{a_q^2 + db_q^2}{4},$$
$$(a_p + 2)^2 + db_p^2 = a_q^2 + db_q^2,$$

from which it follows that $b_q = b_p$. $\qquad\square$

The corresponding case of the above Theorem for $d = *$ will be covered in Subsection 4.2.2. As we shall remark below, the number A_{pq} treated in the following three results is very useful in analyzing the relation $(p, q)_d$.

Theorem 4.2.9. *Let $(p, q)_*$ be an elliptic pair. There exists an integer $A = A_{pq}$ such that*

$$A^2 = \frac{2pq + 2p + 2q - p^2 - q^2 - 1}{3}.$$

Choose the sign on A such that $A \equiv_4 p + q + 1$. Then $E : y^2 = x^3 + g^m$, where g is a particular primitive root modulo q, can have one of six orders over $\mathbb{F}_q$:

1. *If $m \equiv_6 0$: $\#\tilde{E}(\mathbb{F}_q) = \frac{1}{2}(p + q + 1 + 3A)$,*

2. If $m \equiv_6 1$: $\#\tilde{E}(\mathbb{F}_q) = p$,

3. If $m \equiv_6 2$: $\#\tilde{E}(\mathbb{F}_q) = \frac{1}{2}(p + q + 1 - 3A)$,

4. If $m \equiv_6 3$: $\#\tilde{E}(\mathbb{F}_q) = \frac{1}{2}(-p + 3q + 3 - 3A)$,

5. If $m \equiv_6 4$: $\#\tilde{E}(\mathbb{F}_q) = 2q + 2 - p$,

6. If $m \equiv_6 5$: $\#\tilde{E}(\mathbb{F}_q) = \frac{1}{2}(-p + 3q + 3 + 3A)$.

Proof. The existence of A is proven in Theorem 4.1.7(a). The rest follows from comparing the orders in terms of A (found in Theorem 4.1.7(b)) and in terms of the values given in our Theorem 4.2.4. Note that if our choice of primitive root g does not make the order of $E : y^2 = x^3 + g$ over $\mathbb{F}_q$ equal to p, then we can take $g \leftarrow g^{-1}$ to produce the desired result. $\qquad\square$

Corollary 4.2.10. *Let $(p, q)_*$ be an elliptic pair. If we write $p = a^2 + 3b^2$, with $a \equiv_3 -1$, and we choose the sign on b such that $q = p + 1 + a - 3b$, then $A_{pq} = a + b$.*

Proof. A straightforward computation, plugging in $p = a^2 + 3b^2$ and $q = a^2 + 3b^2 + 1 + a - 3b$, shows that $A_{pq}^2 = (a+b)^2$. Then, $A_{pq} \equiv_4 p + q + 1 \equiv_4 p + (p + 1 + a - 3b) + 1 \equiv_4 2p + 2 + a - 3b$. As p is odd, $4 | (2p + 2)$, so $A \equiv_4 a - 3b \equiv_4 a + b$. We know that $a + b$ is odd because p is odd, so $-(a + b) \not\equiv_4 a + b$. Therefore, $A = a + b$. $\qquad\square$

We can also state Corollary 4.2.10 for $d \neq *$.

Corollary 4.2.11. *Let $(p, q)_d$ be an elliptic pair over $d \neq *$ with $p \neq q$. Then there exists an integer A_{pq} such that*

$$A_{pq}^2 = \frac{2pq + 2p + 2q - p^2 - q^2 - 1}{d}. \tag{4.8}$$

Indeed, $A_{pq} = b_p(= b_q)$.

Proof. We write $4p = a_p^2 + db_p^2$. Using the fact that $p < q$ and $(p, q)_d$ is an elliptic pair over d, we have $q = p + 1 + a_p$. A straightforward computation shows that $A = b_p = b_q$. $\qquad\square$

The number A_{pq} is useful: given p, q, and d, we can compute $\frac{2pq+2p+2q-p^2-q^2-1}{d}$, and then we know immediately that if this number is not a perfect square, then $(p, q)_d$ is not an elliptic pair. Moreover, given only p and q within each other's Hasse intervals, we can compute the numerator of the quantity in Equation (4.8) and factor out any perfect squares to compute a d for which $(p, q)_d$ could be an elliptic pair. This provides a convenient proof that p and q sufficiently close together form an elliptic pair for at most one square-free value of d, depending on whether the appropriate elliptic curve is defined modulo p and q.

4.2.2 Elliptic Lists and Elliptic Cycles

Definition 4.2.12. *The symbol $(p_1, p_2, ...p_n)_d$ denotes an **elliptic list of length** n **over** d if each of $(p_1, p_2)_d$, $(p_2, p_3)_d$, ..., $(p_{n-1}, p_n)_d$ is an elliptic pair. If $p_1, p_2, ..., p_n$ are all distinct primes, then the elliptic list is a **proper elliptic list of length** n **over** d.*

Theorem 4.2.13. *Let $(p_1, p_2, p_3, p_4, p_5)_*$ be an elliptic list with $p_1, p_5 \neq p_3$ and $p_2 \neq p_4$, then*

$$p_1 - p_2 = p_5 - p_4.$$

This theorem also holds when $p_1 = p_3 = p_5$ and $p_2 = p_4$, in which case (p, q) is an "amicable pair." The particular form given in Theorem 4.2.13 will be necessary to prove Theorem 4.2.15, however.

Proof. Let $p_3 = a^2 + 3b^2$, and choose b so that $A_{p_3 p_4} = a + b$ (Corollary 4.2.10). Then we apply Theorem 4.2.4 to p_4 to compute that $p_5 = p_3 + 3 + 3a - 3b$. Likewise, $A_{p_2 p_3} = a - b$, so $p_1 = p_3 + 3 + 3a + 3b$. Thus, $p_1 - p_5 = 6b$. By Theorem 4.2.4, $p_2 - p_4 = 6b$, so $p_1 - p_5 = p_2 - p_4$. Rearrangement of terms gives the stated result.

Alternatively, the orders can be read off of Corollary 4.2.17, and this result computed directly. $\square$

Definition 4.2.14. *Let $(p_1, p_2, ...p_n)_d$ form an **elliptic n-cycle over** d (or **elliptic cycle of length** n **over** d) if $(p_1, p_2, p_3, ..., p_n)_d$ forms an elliptic list and $(p_n, p_1)_d$ forms an elliptic pair. If the list is proper, then we say that we have a **proper elliptic n-cycle**.*

Theorem 4.2.15. *If a proper elliptic n-cycle over d exists for $n \geq 3$, then $d = *$ and $n = 6$.*

Proof. When $d \neq *$, this result strengthens Corollary 6.2 of [15], although it utilizes the exact same proof. Recall that if $j \neq 0$, then $\#\tilde{E}(\mathbb{F}_p) = p + 1 \pm a$. In particular, one of these values is greater than p, and the other is no larger than p. If we have a proper elliptic cycle over d, it has a least element. Let p be this least element, and let $(p, q)_d$ and $(r, p)_d$ be the elliptic pairs it is part of. Then either q or r is less than p, a contradiction.

Now let $d = *$. We note that by Theorem 4.2.13, $p_1 - p_2 = p_5 - p_4$. Suppose that $p_1 = p_4$, so that we have a 3-cycle. Then

$$p_5 - p_4 = p_1 - p_2 = p_4 - p_5,$$

so $p_4 = p_5$, and the cycle is not proper.

Suppose instead that $n = 4$. then $p_1 - p_2 = p_5 - p_4$. But because we have a 4-cycle, $p_1 = p_5$, so $p_2 = p_4$, and the cycle is not proper.

Now we let $n = 5$. We have that $p_1 - p_2 = p_5 - p_4$ and $p_2 - p_3 = p_6 - p_5$, so

$$p_6 - p_4 = p_1 - p_3.$$

Setting $p_1 = p_6$ implies that $p_3 = p_4$, so the cycle is not proper.

Now let $n \geq 6$. Again, $p_1 - p_2 = p_5 - p_4$ and $p_2 - p_3 = p_6 - p_5$, so

$$p_6 - p_4 = p_1 - p_3.$$

But now, $p_3 - p_4 = p_7 - p_6$, so

$$p_1 - p_4 = p_7 - p_4,$$

and thus $p_1 = p_7$. This means that if a proper n-cycle has length at least 6, then it has length exactly equal to 6. $\qquad\square$

Given only a single case left to check to see if there exist any proper elliptic cycles other than elliptic pairs (and cycles of length $n = 1$ - which arise from the anomalous primes), we looked for elliptic cycles over $d = *$, using the next three results to narrow our search:

Theorem 4.2.16. *Let $(p_1, p_2, p_3, p_4)_*$ be a proper elliptic list. If $p_1 = a^2 + 3b^2$, with $a \equiv_3 -1$, then $p_4 = (-a - 2)^2 + 3b^2$.*

Proof. See [2]. Note that the proof is purely computational and follows from Corollary 4.2.10. $\qquad\square$

Corollary 4.2.17. *Let $(p_1, p_2, p_3, p_4, p_5, p_6)_*$ be a proper elliptic list, and let $p_1 = a^2 + 3b^2$, with $a \equiv_3 -1$. Then:*

$$p_2 = \left(\frac{a + 3b - 1}{2}\right)^2 + 3\left(\frac{a - b + 1}{2}\right)^2,$$

$$p_3 = \left(\frac{-a + 3b - 3}{2}\right)^2 + 3\left(\frac{a + b + 1}{2}\right)^2,$$

$$p_4 = (-a - 2)^2 + 3b^2,$$

$$p_5 = \left(\frac{-a - 3b - 3}{2}\right)^2 + 3\left(\frac{a - b + 1}{2}\right)^2,$$

$$p_6 = \left(\frac{a - 3b - 1}{2}\right)^2 + 3\left(\frac{a + b + 1}{2}\right)^2.$$

Proof. The proof is again computational, with the orders being the same as the ones found in the proof of Theorem 4.2.16 (see [2]). $\qquad\square$

Theorem 4.2.18. *If $p = a^2 + 3b^2$ is part of a proper elliptic 6-cycle over $d = *$, then $a \equiv_7 -1$ and $7|b$.*

Proof. The proof is entirely computational, taking the orders found in Corollary 4.2.17 and then computing all the primes modulo 7. We find that $a \equiv_7 -1$ and $7|b$, or else at least one prime in the cycle is divisible by 7. The cases are given in Table A.1 in Appendix A.1. $\qquad\square$

Due to the structure imposed by Corollary 4.2.17, it is clear that we can determine all the primes in a proper elliptic 6-cycle given a single prime in the cycle. Using divisibility by 7 to eliminate possible cases, we wrote a computer program to search the remaining values of a and b to find 6-cycles. It found them almost immediately, with the smallest being

$$(275269, 274723, 275227, 276277, 276823, 276319)_*,$$

corresponding to $(a, b) = (251, 266)$ (see Corollary 4.2.17 for notation).

The concept of a proper elliptic 6-cycle is similar to the notion of an **aliquot cycle** defined by Silverman and Stange [15], except that an aliquot cycle fixes a curve E (defined over $L = \mathbb{Q}$, although this can easily be generalized to arbitrary extension of $\mathbb{Q}$) such that $\#\tilde{E}(\mathbb{F}_{p_1}) = p_2, \#\tilde{E}(\mathbb{F}_{p_2}) = p_3, ..., \#\tilde{E}(\mathbb{F}_{p_n}) = p_1$. It is not too difficult to see that any proper elliptic cycle can be made into an aliquot cycle. Find curves $E_i/\mathbb{F}_{p_i}$, $1 \leq i \leq n$ such that $\#E_i(\mathbb{F}_{p_i}) = p_{i+1}$ for $1 \leq i \leq n-1$ and $\#E_i(\mathbb{F}_{p_n}) = p_1$. The coefficients of E_i are only defined uniquely modulo p_i, so we can use the Chinese Remainder Theorem to find the unique coefficients modulo $\prod_{i=1}^{n} p_i$ which are equivalent to the coefficients of E_i modulo p_i for all $1 \leq i \leq n$. In the case of the list above, we can rewrite it as

$$(274723, 275269, 276319, 276823, 276277, 275227)_*$$

to get the smallest (normalized) aliquot cycle corresponding to the curve $E : y^2 = x^3 + 15$.

In the course of computing the cycle above, we also found other 6-cycles with 3 primes represented, each twice. Here, $p_1 = p_2$, $p_4 = p_5$, and $p_3 = p_6$, so that p_1 and p_4 are the so-called **anomalous primes**. We found three cycles of this form with primes less than one million. They are:

$$(114661, 114661, 115249, 115837, 115837, 115249)_*,$$
$$(169219, 169219, 169933, 170647, 170647, 169933)_*,$$
$$(283669, 283669, 284593, 285517, 285517, 284593)_*.$$

After these cycles, the next smallest one has primes greater than ten million.

4.2.3   Properties of Proper Elliptic Lists

We saw in Subsection 4.2.2 that proper elliptic lists over $d = *$ have length no longer than six. In this section, we explore proper elliptic lists over $d \neq 3$. Such lists are either increasing or decreasing (because $\#\tilde{E}(\mathbb{F}_p) = p + 1 \pm a$; see the proof of Theorem 4.2.15), so we will assume throughout that lists are written in ascending order.

Theorem 4.2.8 allows us to describe any proper elliptic list over $d \neq *$ if we have just one of its members. The following Theorem is an extension of Theorem 4.2.8, and follows immediately from it and the definition of an elliptic list:

Theorem 4.2.19. *Let $(p_1, \ldots, p_n)_d$ be a proper elliptic list over $d \neq *$, and let $a_{p_1}, \ldots, a_{p_n}$ and $b_{p_1}, \ldots, b_{p_n}$ be the unique positive integers such that $4p_i = a_{p_i}^2 + db_{p_i}^2$ for each $i = 1, \ldots, n$. Then $a_{p_i} = a_{p_1} + 2i - 2$ and $b_{p_i} = b_{p_1}$ for each $i = 1, \ldots, n$.*

Theorem 4.2.20. *Let $d \neq 3$ be a square-free positive integer. Consider a prime number p_1 which is of the form $4p_1 = a^2 + db^2$. If p_1 is the initial term of a proper elliptic list of length n over d, then the quadratic polynomial $x^2 + ax + p_1$ has n consecutive prime values for $x = 0, 1, \ldots, n-1$.*

Theorem 4.2.19 implies that any proper elliptic list $(p_1, \ldots, p_n)_d$ over $d \neq *$ is generated by n consecutive prime values of the quadratic polynomial in i

$$\frac{1}{4}((2i + a_1)^2 + db_1^2)$$
$$= i^2 + a_1 i + p_1$$

for $i = 0, 1, \ldots, n-1$. Conversely, any such polynomial will generate an elliptic list. Therefore, the problem of finding proper elliptic lists over $d \neq *$ is equivalent to finding prime-generating polynomials of the form above.

The longest proper list we've found so far is:

$$(41, 43, 47, 53, 61, 71, 83, 97, 113, 131, 151, 173, 197, 223, 251, 281, 313,$$
$$347, 383, 421, 461, 503, 547, 593, 641, 691, 743, 797, 853, 911,$$
$$971, 1033, 1097, 1163, 1231, 1301, 1373, 1447, 1523, 1601)_{163}.$$

This list corresponds to the famous polynomial $n^2 + n + 41$, which is prime for $n = 0, 1, \ldots, 39$.

It is still unclear whether proper elliptic lists of arbitrarily long length exist, although we have an upper bound for the length given d in Theorem 4.2.21.

Theorem 4.2.21. *Let $d \equiv_8 3$, $d \neq *$. Let z be the smallest prime such that $\left(\frac{-d}{z}\right) \neq -1$. Then there are no proper elliptic lists over d of length $n \geq z$.*

Proof. Let $(p_1, \ldots, p_n)_d$ be an elliptic list of length n. Let a_i, b_i be the unique positive integers such that $4p_i = a_i^2 + db_i^2$, and note that for all $1 \leq i \leq n$, $a_i = a_1 + 2i - 2$, $b_i = b_1 = b$ (Theorem 4.2.8).

First consider the case $\left(\frac{-d}{z}\right) = 0$. Assume that $n \geq z$. The numbers $a_1, a_2, \ldots, a_n$ take every value modulo z because z is odd and the a_i's increment by two, so one of them, say a_j, will be divisible by z. Therefore,

$$4p_j = a_j^2 + db^2 \equiv_z 0 + 0 \cdot b \equiv_z 0,$$

and thus $z | p_j$, a contradiction unless $z = p_j$, which we cover below.

Now consider the case $\left(\frac{-d}{z}\right) = 1$. Assume that $n \geq z$. As before, the numbers $a_1, a_2, \ldots, a_n$ take every value modulo z, so one of them, say a_j, will satisfy $a_j \equiv_z b\sqrt{-d}$. Therefore,

$$4p_j = a_j^2 + db^2 \equiv_z -db^2 + db^2 \equiv_z 0,$$

implying that $z | p_j$, a contradiction unless $z = p_j$, which we cover below.

It remains only to consider the case that one of the elements in the cycle is in fact z. The smallest number which can be written as $\frac{1}{4}(a^2 + db^2)$ for $a, b \in \mathbb{N}$ is at least $\frac{d+1}{4}$, and z is never more than this value, so we only have to look at the case $p_1 = z = \frac{d+1}{4}$. In this case, $a_1 = b_1 = 1$, so $p_i = \frac{(2i-1)^2 + d}{4}$, and $p_z = \frac{(2z-1)^2 + (4z-1)}{4} = z^2$ is composite. $\qquad\square$

44

In particular, we see that for $d = 3$, $\left(\frac{-d}{3}\right) = 0$, so we cannot construct any proper elliptic lists of length $n \geq 3$ in this case - only elliptic pairs.

Corollary 4.2.22. *Assume that $d \equiv_8 3$, and $d \neq *$. If $d \not\equiv_{24} 19$, then there do not exist any proper elliptic lists of length $n \geq 3$ over d.*

Proof. By hypothesis, $d \equiv_{24} 3$ or 11. In the first case, $\left(\frac{-d}{3}\right) = 0$, and in the latter case, $\left(\frac{-d}{3}\right) = 1$. □

Definition 4.2.23. *Restrict d to be a positive and square-free integer.*

$$\mathcal{M}(d) = \begin{cases} \min\{z > 1 : z \text{ prime and } \left(\frac{-d}{z}\right) \neq -1\} & if\ d \equiv_8 3 \\ 1 & otherwise \end{cases}$$

$\mathcal{M}(d)$ *is the **maximum allowable list-length function**.*

For $d \not\equiv_8 3$, there are no lists of any length, motivating our definition of $\mathcal{M}(d) = 1$. For $d \equiv_8 3$, the value of $\mathcal{M}(d)$ is motivated by Theorem 4.2.21. Note that in the case $d = *$, there exist proper elliptic lists of length up to 6, so $\mathcal{M}(*) = 7$.

Currently we have no reason to believe that a list of length $\mathcal{M}(d) - 1$ exists over d (except in the case that the class number $h(K) = 1$), motivating our next definition:

Definition 4.2.24. *Restrict d to be a positive and square-free integer. Define*

$$\mathcal{L}(d) = \max\{n | (p_1, ..., p_n)_d \text{ is a proper elliptic list}\}.$$

Also define

$$f(d) = \mathcal{M}(d) - \mathcal{L}(d).$$

We have $\mathcal{L}(d) < \mathcal{M}(d)$ and $f(d) > 0$. We suspect that $f(d)$ is related to the class number $h(-d) = h(K)$ in some way, but we have not been able to prove any results so far, other than the fact that $f(d) = 1$ if $h(-d) = 1$.

4.2.4 The Relationship between Elliptic Pairs and Current Conjectures

We now lump $d = *$ and $d = 3$ into a single case, since they take on the same elliptic pairs (even though $d = *$ admits more elliptic lists and cycles).

The distribution of elliptic pairs is of primary importance because it determines the time and space complexity of algorithms for generating elliptic curves of prime order (see [4], for example). Although thus far nobody has explicitly formulated a conjecture for this, many similar heuristics and conjectures (see [4],[7],[15],[19]) suggest that the number of elliptic pairs $(p, q)_d$ with primes $p \leq q$ less than X should be asymptotic to

$$C_d \frac{X}{log^2 X}$$

for some constant C_d as $X \to \infty$. It is believed that C_d is greater than zero for all positive square-free integers $d \equiv_8 3$, but so far it is not even known if the number of elliptic pairs for any given d is infinite.

The remark at the end of Section 4.2.1 indicates that for every pair of primes within each other's Hasse interval, we have an elliptic pair. By the prime number theorem, for any given p, the probability that another number q near it is prime is approximately $1/\log p$, so p forms an elliptic pair with about $4\sqrt{p}/\log p$ primes, of which roughly $2\sqrt{p}/\log p$ are greater than p (so that we only consider unique elliptic pairs, since Theorem 4.2.7 indicates that the order of p and q does not matter). Since $d \leq 4p - 1$ and $d \equiv_8 3$, we have roughly $p/2$ values of d which could make $(p,q)_d$ an elliptic pair, but each square-free d is equivalent to $\lfloor \frac{1}{2}\sqrt{\frac{p}{d}} \rfloor$ of these values (obtained by multiplying d by the squares of odd numbers), so we'd expect any particular d to be chosen with probability about $(2\sqrt{p}/\log p)/\left(\frac{p}{2} / \left(\frac{1}{2}\sqrt{\frac{p}{d}} \right) \right) = 2/\sqrt{d}\log p$. Since $\pi(X) \sim \frac{X}{\log X}$, this suggests to us that for any given d, the number of elliptic pairs with primes less than sufficiently large X should be about

$$\sum_{p \leq X} \frac{2}{\sqrt{d}\log p} \sim \sum_{n=1}^{X/\log X} \frac{2}{\sqrt{d}\log(n \log n)} \sim \sum_{n=1}^{X/\log X} \frac{2}{\sqrt{d}\log n} \asymp \frac{1}{\sqrt{d}} \frac{X}{\log^2 X}.$$

Because the above sum diverges as $X \to \infty$, we see that for at least one value of d, there must be an infinite number of elliptic pairs. Furthermore, we would expect that $C_d/C_{d'} \approx \sqrt{d'/d}$ as a first-order approximation. Unfortunately, this contradicts experimental data obtained by Silverman and Stange in Section 9 of [15], which shows a reciprocal relationship for d's with class number $h(-d) = 1$. As $-d$ grows larger, however, the frequency of elliptic pairs decreases in general. Our (implicit) assumption of a uniform distribution of elliptic primes among the d's must therefore be false, so our heuristic must be altered in some way. In his proof of the prime number theorem in 1896, de la Vallée Poussin showed that the number of primes less than X represented by a binary quadratic form Q of discriminant $-d$ is $\pi_Q(X) = \frac{1}{2h(-d)}\mathrm{Li}(X) + O(X\exp(-c_Q\sqrt{\log X}))$, where $\mathrm{Li}(X)$ is the usual logarithmic integral [17]. Since $(p,q)_d$ corresponds to a quadratic form of discriminant $-d$, p and q are each chosen from a set of primes of Dirichlet density $\frac{1}{2h(-d)}$. Therefore, we expect the number of elliptic pairs given d to be proportional to $\frac{1}{[h(-d)]^2}$ (since we have two primes). In order to have the correct number of primes, however, we need $C_d \asymp \frac{1}{\sqrt{d}} \asymp \frac{n(d)}{[h(-d)]^2}$. As a first-order approximation, $n(d) \approx \sqrt{d}$, so $C_d \sim \frac{\sqrt{d}}{[h(-d)]^2}$, gives very similar results to the data we collected. Higher order terms probably replace $n(d)$ by $n(d)\log^k d$ for $k = 1, 2, 3, ...$, although we have not performed any numerical tests to verify this.

We computed d for every possible pair of primes (p,q), each less than 10^7, and examined the results for all $d \equiv_8 3$ such that $h(-d) = 1, 2, 3, 4$, neglecting the possibility that the representative curve might not be defined (which we would expect to change the results by a factor depending on h). We found that $C_d \sim C\frac{\sqrt{d}}{[h(-d)]^2}$, in agreement with our guess, with $C \approx 0.16$. Unfortunately, our results are ineffective in computing the actual value of C_d, since we currently do not know how to find the higher-order terms, although it may be possible

46

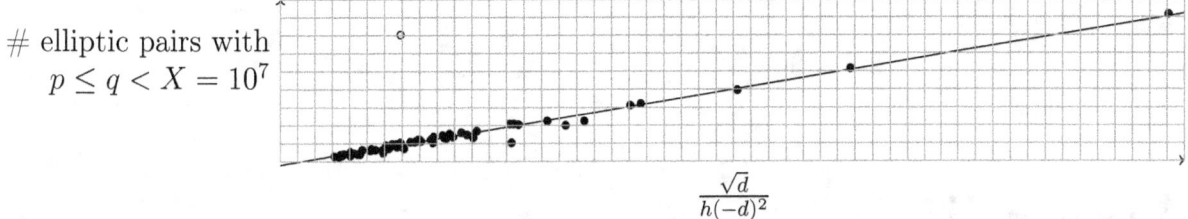

Figure 4.1: Plot of the number of elliptic pairs $(p, q)_d$ with $p, q < X = 10^7$ divided by $\frac{X}{\log^2 X}$ as a function of $\frac{\sqrt{d}}{[h(-d)]^2}$ for $d \equiv_8 3$ and $h(-d) \leq 4$ (neglecting the possibility that all representative curves might be undefined over $\mathbb{F}_p$ or $\mathbb{F}_q$). Note that we had to treat the $d = 3$ case (open diamond) separate from the others, as always. The linear fit suggests that our asymptotic relation, $C_d \asymp \frac{\sqrt{d}}{[h(-d)]^2}$ is reasonable.

to compute $n(d)$ for small d. Also, the value $C_3 \approx 1.757$ is about 6.26 times greater than expected, so it is likely that the $d = 3$ case must be examined using a different heuristic, just as it had to be treated separately earlier. Dirichlet's class number formula requires division by an extra factor of 3 for $d = 3$, so this could potentially account for the discrepancy, along with the fact that $n(d)$ appears to grow a little faster than $\sqrt{d}$. Our results are given in Figure 4.1, with the data listed in Appendix A.2.

Elliptic pairs are also related to the Bouniakowsky conjecture for quadratic polynomials. The Bouniakowsky conjecture states that all irreducible polynomials $p(x)$ with integer coefficients of degree greater than or equal to two such that there does not exist a prime m which divides $p(x)$ for all integers x take on an infinite number of prime values. Such polynomials are called **Bouniakowsky polynomials**. The case for polynomials of degree one was proven by Dirichlet.

Theorem 4.2.25. *The Bouniakowsky conjecture (for quadratic polynomials) implies that there are an infinite number of anomalous primes (and thus elliptic pairs) over any $d \equiv_8 3$.*

Proof. Let $d = *$, and let $(p, p)_*$ be an elliptic pair. Then $p = p + 1 + a - 3b$, so $a = 3b - 1$. As $p = a^2 + 3b^2$, we have that

$$p(b) = (3b - 1)^2 + 3b^2 = 12b^2 - 6b + 1,$$

which is easily checked to be a Bouniakowsky polynomial.

If $d \neq *$, then $(p, p)_d$ is an elliptic pair if and only if $p = p + 1 - a$, or $a = 1$. Thus, $p = \frac{1}{4}(1 + db^2)$. We know that b must be odd so that p is an integer, so we set $b = 2c + 1$ and then

$$p(c) = \frac{1}{4}(1 + d(2c + 1)^2) = dc^2 + dc + \frac{1 + d}{4}.$$

Since $4 | (1 + d)$, this polynomial has integer coefficients, and it is easy to check that it is a Bouniakowsky polynomial. $\qquad\square$

Appendix A

Supplementary Information for Section 4.2

A.1   Proof of Theorem 4.2.18

Let $p_1 = a^2 + 3b^2$, with $a \equiv_3 -1$ and the sign on b chosen such that $p_2 = p_1 + 1 + a - 3b$, and assume that $(p_1, p_2, p_3, p_4, p_5, p_6)_*$ is a proper elliptic cycle. By Corollary 4.2.17, a and b determine the cycle uniquely. We enumerate over all the possible choices of a and b modulo 7 in Table A.1. In the last column, we take $\prod\limits_{i=1}^{6} p_i \mod 7$. If this is 0, then at least one p_i is not prime (since 7 is not part of a proper elliptic cycle of length 6). Only when $a \equiv_7 6 \equiv_7 -1$ and $b \equiv_7 0$ do we get a 1 in the last column and the possibility of a non-trivial proper elliptic cycle.

A.2   Data for Asymptotic Evaluation of C_d

We conjecture that the number of elliptic pairs $(p, q)_d$ with $p \leq q$ less than some given upper bound X is asymptotic to $C_d \frac{X}{\log^2 X}$ as $X \to \infty$, with $C_d \asymp \frac{\sqrt{d}}{[h(-d)^2]^2}$ as $C_d \to \infty$. Of course, C_d is bounded, and it is frequently very small, so this relation is computationally ineffective. The data, shown in Figure 4.1, support this conjecture, however. The rest of this appendix consists of a table (Table A.2) of the data we collected for values of d with class number $h(-d) \leq 4$.

Table A.1: Primes in Elliptic Cycles Modulo 7

a mod 7	b mod 7	p_1 mod 7	p_2 mod 7	p_3 mod 7	p_4 mod 7	p_5 mod 7	p_6 mod 7	$\prod_{i=1}^{6} p_i$ mod 7
0	0	0	1	3	4	3	1	0
0	1	3	1	3	0	2	0	0
0	2	5	0	2	2	0	5	0
0	3	6	5	0	3	4	2	0
0	4	6	2	4	3	0	5	0
0	5	5	5	0	2	2	0	0
0	6	3	0	2	0	3	1	0
1	0	1	3	0	2	0	3	0
1	1	4	3	0	5	6	2	0
1	2	6	2	6	0	4	0	0
1	3	0	0	4	1	1	4	0
1	4	0	4	1	1	4	0	0
1	5	6	0	4	0	6	2	0
1	6	4	2	6	5	0	3	0
2	0	4	0	6	2	6	0	0
2	1	0	0	6	5	5	6	0
2	2	2	6	5	0	3	4	0
2	3	3	4	3	1	0	1	0
2	4	3	1	0	1	3	4	0
2	5	2	4	3	0	5	6	0
2	6	0	6	5	5	6	0	0
3	0	2	6	0	4	0	6	0
3	1	5	6	0	0	6	5	0
3	2	0	5	6	2	4	3	0
3	3	1	3	4	3	1	0	0
3	4	1	0	1	3	4	3	0
3	5	0	3	4	2	6	5	0
3	6	5	5	6	0	0	6	0
4	0	2	0	3	1	3	0	0
4	1	5	0	3	4	2	6	0
4	2	0	6	2	6	0	4	0
4	3	1	4	0	0	4	1	0
4	4	1	1	4	0	0	4	0
4	5	0	4	0	6	2	6	0
4	6	5	6	2	4	3	0	0
5	0	4	3	1	0	1	3	0
5	1	0	3	1	3	0	2	0
5	2	2	2	0	5	5	0	0
5	3	3	0	5	6	2	4	0
5	4	3	4	2	6	5	0	0
5	5	2	0	5	5	0	2	0
5	6	0	2	0	3	1	3	0
6	**0**	**1**	**1**	**1**	**1**	**1**	**1**	**1**
6	1	4	1	1	4	0	0	0
6	2	6	0	0	6	5	5	0
6	3	0	5	5	0	2	2	0
6	4	0	2	2	0	5	5	0
6	5	6	5	5	6	0	0	0
6	6	4	0	0	4	1	1	0

Table A.2: $Y =$ the Number of Elliptic Pairs $(p, q)_d$ with $p \le q < X = 10^7$

d	$h(-d)$	$\frac{\sqrt{d}}{[h(-d)]^2}$	Y	$\frac{Y}{X/\log^2 X}$
3	1	1.732051	67619	1.756694
11	1	3.316625	10125	0.263040
19	1	4.358899	21466	0.557672
43	1	6.557439	38158	0.991318
67	1	8.185353	49662	1.290184
163	1	12.76715	78517	2.039817
35	2	1.479020	4545	0.118076
51	2	1.785357	7054	0.183258
91	2	2.384848	12324	0.320169
115	2	2.680951	14274	0.370829
123	2	2.772634	12669	0.329132
187	2	3.418699	19490	0.506337
235	2	3.832427	21643	0.562270
267	2	4.085034	19062	0.495217
403	2	5.018715	29974	0.778704
427	2	5.165995	30647	0.796188
59	3	0.853461	2456	0.063805
83	3	1.012270	3238	0.084121
107	3	1.149342	3845	0.099890
139	3	1.309981	6503	0.168943
211	3	1.613982	8458	0.219733
283	3	1.869178	10504	0.272887
307	3	1.946824	10947	0.284395
331	3	2.021489	10676	0.277355
379	3	2.163102	11513	0.299100
499	3	2.482034	13267	0.344667
547	3	2.598670	15323	0.398081
643	3	2.817494	16295	0.423333
883	3	3.301702	20009	0.519820
907	3	3.346271	19969	0.518781
155	4	0.778119	2541	0.066013
195	4	0.872765	3510	0.091187
203	4	0.890488	3019	0.078432
219	4	0.924916	3678	0.095552
259	4	1.005842	5135	0.133404
291	4	1.066170	4255	0.110542
323	4	1.123263	3944	0.102462
355	4	1.177590	6290	0.163410
435	4	1.303541	5661	0.147069
483	4	1.373579	6076	0.157850
555	4	1.472402	6347	0.164891
595	4	1.524539	8283	0.215187
627	4	1.564998	7084	0.184037
667	4	1.614146	9325	0.242257
715	4	1.671218	9364	0.243270
723	4	1.680541	7687	0.199703
763	4	1.726403	10009	0.260027
795	4	1.762234	7752	0.201392
955	4	1.931442	10565	0.274471
1003	4	1.979386	11738	0.304945
1027	4	2.002928	11536	0.299697
1227	4	2.189285	10072	0.261664
1243	4	2.203513	13164	0.341992
1387	4	2.327653	14006	0.363866
1411	4	2.347705	12649	0.328612
1435	4	2.367587	13188	0.342615
1507	4	2.426256	14544	0.377843
1555	4	2.464593	14049	0.364983

Appendix B

Code used to Generate Plots of Elliptic Curves

```cpp
// ECTC.cpp
// Created 19-20 December 2012 by Thomas Morrell

#include <cstdlib>
#include <iostream>
#include <fstream>
#include <cmath>

using namespace std;

int main(int argc, char *argv[])
{
    cout << "y^2 = x^3 + Ax + B" << endl;
    float a,b;
    cout << "A = ";
    cin >> a;
    cout << "B = ";
    cin >> b;

    cout << endl << "RANGES:" << endl;
    float xMin, xMax, yMin, yMax, rate;
    cout << "x_min = ";
    cin >> xMin;
    cout << "x_max = ";
    cin >> xMax;
    cout << "For y-range to function properly, set y_min < 0, y_max > 0\n";
    cout << "y_min = ";
    cin >> yMin;
    cout << "y_max = ";
```

```cpp
cin >> yMax;
// Sample rate is how often points are found with respect to x.
cout << "Sample rate = ";
cin >> rate;

// Scale transforms x,y into scale*x,scale*y for printing to the file
cout << endl << "Scale = ";
float scale;
cin >> scale;

cout << endl << "FINDING EXTREMA...";
float min1, max1, min2, maxY2;
min1 = xMin-1;
max1 = xMax+1;
min2 = xMin-1;
maxY2 = yMin * yMin;
if (maxY2 < yMax * yMax) { maxY2 = yMax * yMax; }
for (int i = 0; i <= (xMax-xMin)/rate; i++)
{
    float x = xMin + i * rate;
    float y2 = x * x * x + a * x + b;
    if (y2 >= 0.0)
    {
        if (min1 < xMin) { min1 = x; }
        else if (max1 < xMax) { min2 = x; }
    }
    if (y2 < 0)
    {
        if (min1 > xMin) { max1 = x - rate; }
    }
}
if (max1 > xMax) { max1 = xMax; }

cout << endl << "PLOTTING POINTS...";
ofstream myfile;
myfile.open ("EC.table");
myfile << "# y^2 = x^3 + " << a << " * x + " << b << endl;
myfile << "# x y Comments" << endl;
int size = int((max1-min1)/rate) + 1;
float yVals[size];
for (int i = 0; i < size; i++)
{
    float x = min1 + i * rate;
    float y2 = x * x * x + a * x + b;
    yVals[i] = sqrt(y2);
```

```cpp
}
for (int i = 0; i < size; i++)
{
    if (yVals[size - 1 - i] <= -yMin)
    {
        myfile << (min1 + (size - 1 - i) * rate) * scale << " "
               << 0 - scale * yVals[size - 1 - i] << endl;
    }
}
for (int i = 0; i < size; i++)
{
    if (yVals[i] <= yMax)
    {
        myfile << (min1 + i * rate) * scale << " "
               << scale * yVals[i] << endl;
    }
}
myfile.close();

cout << endl << "SECOND SEGMENT?...";
if (min2 > xMin)
{
    ofstream myfile2;
    myfile2.open ("EC2.table");
    myfile2 << "# y^2 = x^3 + " << a << " * x + " << b << endl;
    myfile2 << "# Second segment" << endl;
    myfile2 << "# x y Comments" << endl;
    size = int((xMax-min2)/rate) + 1;
    float yVals[size];
    for (int i = 0; i < size; i++)
    {
        float x = min2 + i * rate;
        float y2 = x * x * x + a * x + b;
        yVals[i] = sqrt(y2);
    }
    for (int i = 0; i < size; i++)
    {
        if (yVals[size - 1 - i] <= -yMin)
        {
            myfile2 << (min2 + (size - 1 - i) * rate) * scale << " "
                    << 0 - scale * yVals[size - 1 - i] << endl;
        }
    }
    for (int i = 0; i < size; i++)
    {
```

```cpp
            if (yVals[i] <= yMax)
            {
                myfile2 << (min2 + i * rate) * scale << " "
                        << scale * yVals[i] << endl;
            }
        }
        myfile2.close();
    }

    cout << endl << "GENERATING TikZ CODE...";
    ofstream myfile3;
    myfile3.open ("TikZ.txt");
    myfile3 << "\\begin{center}\n\\begin{figure}\n"
            << "\\begin{tikzpicture}[domain=" << xMin * scale - rate
            << ":" << xMax * scale + rate << "]\n"
            << "\\draw[->] (" << xMin * scale - rate << ",0) -- ("
            << xMax * scale + rate << ",0) node[right] {$x$};\n"
            << "\\draw[->] (0," << yMin * scale - rate << ") -- (0,"
            << yMax * scale + rate << ") node[above] {$y$};\n"
            << "\\draw plot[smooth] file {EC.table};"
            << "% MAY NEED TO CHANGE!!!\n";
    if (min2 > xMin)
    {
        myfile3 << "\\draw plot[smooth] file {EC2.table};"
                << "% MAY NEED TO CHANGE!!!\n";
    }
    myfile3 << "\\end{tikzpicture}\n" << "\\caption{$E : y^2 = x^3 + "
            << a << "x + " << b << "$} % MAY NEED TO CHANGE!!!\n"
            << "\\end{figure}\n\\end{center}\n";
    myfile3.close();

    cout << endl << "FINISHED!" << endl;

    system("PAUSE");
    return EXIT_SUCCESS;
}
```

Bibliography

[1] A. O. L. Atkin and F. Morain, *Elliptic curves and primality proving*, **Mathematics of Computation** 61 (1993), 29–68.

[2] L. Babinkostova, K. Bombardier, M. Cole, T. Morrell, C. Scott, *Elliptic Reciprocity*, in preparation, arXiv:1212.1983. (Accessed 23 December 2012.)

[3] B.C. Berndt, R.J. Evans, and K.S. Willams, *Gauss and Jacobi Sums*, vol. 21 of **Canadian Mathematical Society Series of Monographs and Advanced Texts**. Wiley-Interscience, 1998.

[4] R. Bröker and P. Stevenhagen, *Constructing elliptic curves of prime order*, **Contemporary Mathematics** 463 (2008), 17-28.

[5] H. Cohen, *A Course in Computational Algebraic Number Theory*, **Graduate Texts in Mathematics** 138 Springer-Verlag, 1993.

[6] M. Kerr, *Shimura varieties: a Hodge-theoretic perspective*, preprint, available online: http://www.math.wustl.edu/~matkerr/SV.pdf. (Accessed 3 March 2013.)

[7] N. Koblitz, *Primality of the number of points on an elliptic curve over a finite field*, **Pacific Journal of Mathematics**, 131:1 (1988), 157–165.

[8] S. Lang, *Elliptic Functions*, **Graduate Texts in Mathematics** 112 Springer-Verlag, 1987.

[9] R. Schoof, *Elliptic curves over finite fields and the computation of square roots mod p*, **Mathematics of Computation**, 44 (170):483-494, 1985.

[10] M. Scott, ftp://ftp.computing.dcu.ie/pub/crypto/schoof.cpp, 1999. (Accessed 23 December 2012.)

[11] M. Scott, ftp://ftp.computing.dcu.ie/pub/crypto/schoof2.cpp, 2000. (Accessed 23 December 2012.)

[12] A. Silverberg, *Group order formulas for reductions of cm elliptic curves*, **Contemporary Mathematics** 521 (2010), 107–120.

[13] J.H. Silverman, *The Arithmetic of Elliptic Curves*, **Graduate Texts in Mathematics** 106 Springer-Verlag, 1986.

[14] J.H. Silverman, *Advanced Topics in the Arithmetic of Elliptic Curves*, **Graduate Texts in Mathematics** 151 Springer-Verlag, 1994.

[15] J.H. Silverman and K.E. Stange, *Amicable pairs and aliquot cycles for elliptic curves*, **Experimental Mathematics**, 20:3 (2011), 329–357.

[16] J.H. Silverman and J.T. Tate, *Rational Points on Elliptic Curves*, **Undergraduate Texts in Mathematics** Springer-Verlag, 1992.

[17] C.J. de la Vallée Poussin, *Recherches analytiques sur la theorie des nombre premiers*, **Ann. Soc. Sci. Bruxelles**, 20 (1896), 183–256.

[18] L.C. Washington, *Number Theory: Elliptic Curves and Cryptography*, vol. 50 of **Discrete Mathematics and Its Applications**. Chapman & Hall/CRC, 2nd ed., 2008.

[19] D. Zywina, *A refinement of Koblitz's conjecture*, **International Journal of Number Theory**, 7:3 (2011), 739–769.

Index

www.ingramcontent.com/pod-product-compliance
Lightning Source LLC
Chambersburg PA
CBHW081226170526
45165CB00009B/2967

Table of Contents